AF514715

Tucholsky Wagner Zola Scott Sydow Schlegel
Turgenev Wallace Fonatne Freud
Twain Walther von der Vogelweide Fouqué Friedrich II. von Preußen
Weber Freiligrath Frey
Fechner Fichte Weiße Rose von Fallersleben Kant Ernst Richthofen Frommel
Hölderlin
Fehrs Engels Fielding Eichendorff Tacitus Dumas
Faber Flaubert
Maximilian I. von Habsburg Fock Eliasberg Zweig Ebner Eschenbach
Feuerbach Ewald Eliot Vergil
Goethe London
Mendelssohn Balzac Shakespeare Elisabeth von Österreich Dostojewski Ganghofer
Trackl Stevenson Lichtenberg Rathenau Doyle Gjellerup
Mommsen Tolstoi Hambruch
Thoma Lenz Hanrieder Droste-Hülshoff
Dach Verne von Arnim Hägele Hauff Humboldt
Reuter Rousseau Hagen Hauptmann
Karrillon Garschin Gautier
Damaschke Defoe Hebbel Baudelaire
Descartes
Wolfram von Eschenbach Schopenhauer Hegel Kussmaul Herder
Bronner Darwin Melville Dickens Grimm Jerome Rilke George
Campe Horváth Aristoteles Bebel Proust
Bismarck Vigny Gengenbach Barlach Voltaire Federer Herodot
Heine
Storm Casanova Tersteegen Gilm Grillparzer Georgy
Chamberlain Lessing Langbein Gryphius
Brentano Lafontaine
Strachwitz Claudius Schiller Kralik Iffland Sokrates
Katharina II. von Rußland Bellamy Schilling
Gerstäcker Raabe Gibbon Tschechow
Löns Hesse Hoffmann Gogol Wilde Gleim Vulpius
Luther Heym Hofmannsthal Klee Hölty Morgenstern
Roth Heyse Klopstock Kleist Goedicke
Luxemburg Puschkin Homer Mörike
La Roche Horaz Musil
Machiavelli Kierkegaard Kraft Kraus
Navarra Aurel Musset
Lamprecht Kind Kirchhoff Hugo Moltke
Nestroy Marie de France
Nietzsche Nansen Laotse Ipsen Liebknecht
Marx Ringelnatz
von Ossietzky Lassalle Gorki Klett Leibniz
May vom Stein Lawrence Irving
Petalozzi Knigge
Platon Pückler Michelangelo Kafka
Sachs Poe Liebermann Kock
Korolenko
de Sade Praetorius Mistral Zetkin

Forty Centuries of Ink or, a chronological narrative concerning ink and its backgrounds

David Nunes Carvalho

Imprint

This book is part of the TREDITION CLASSICS series.

Author: David Nunes Carvalho
Cover design: toepferschumann, Berlin (Germany)

Publisher: tredition GmbH, Hamburg (Germany)
ISBN: 978-3-8495-2460-9

www.tredition.com
www.tredition.de

The intention of the TREDITION CLASSICS series is to make world literature in the public domain available in printed format. Literary enthusiasts and organizations worldwide have scanned and digitally edited the original texts. tredition has subsequently formatted and redesigned the content into a modern reading layout. Therefore, we cannot guarantee the exact reproduction of the original format of a particular historic edition. Please also note that no modifications have been made to the spelling, therefore it may differ from the orthography used today.

FORTY CENTURIES OF INK

OR

A CHRONOLOGICAL NARRATIVE CONCERNING INK AND ITS BACKGROUNDS

INTRODUCING INCIDENTAL OBSERVATIONS AND DEDUCTIONS, PARALLELS OF TIME AND COLOR PHENOMENA, BIBLIOGRAPHY, CHEMISTRY, POETICAL EFFUSIONS, CITATIONS, ANECDOTES AND CURIOSA TOGETHER WITH SOME EVIDENCE RESPECTING THE EVANESCENT CHARACTER OF MOST INKS OF TO-DAY AND AN EPITOME OF CHEMICO-LEGAL INK.

BY

DAVID N. CARVALHO

PREFACE.

The unfortunate conditions surrounding the almost universal use of the oddly named commercial and with few exceptions record inks, and the so-called modern paper, is the motive for the writing of this book. The numerous color products of coal tar, now so largely employed in the preparation of ink, and the worse material utilized in the manufacture of the hard- finished writing papers, menace the future preservation of public and other records. Those who occupy official position and who can help to ameliorate this increasing evil, should begin to do so without delay. Abroad England, Germany and France and at home Massachusetts and Connecticut have sought to modify these conditions by legislation and our National Treasury Department only last year, in establishing a standard for its ink, gives official recognition of these truths.

There is no "History of Ink;" but of ink history there is a wealth of material, although historians have neglected to record information about the very substance by which they sought to keep and transmit the chronicles they most desired to preserve. From the beginning of the Christian era to the present day, "Ink" literature, exclusive of its etymology, chemical formulas, and methods of manufacture, has been confined to brief statements in the encyclopedias, which but repeat each other. A half dozen original articles, covering only some particular branch together with a few treatises more general in their ramifications of the subject, can also be found. Seventy lines about "writing ink" covering its history for nearly four thousand years is all that is said in "The Origin and Progress of Handwriting," a revised book of hundreds of pages of Sir Thomas Astle, London, 1876, and once deemed the very highest authority.

The mass of ancient and comparatively modern documents which we have inherited, chronicle nothing about the material with which they were written. The more valuable of them are disfigured by the superscription of newer writings over the partially erased earlier ones, thus rendering the work of ascertaining their real character most difficult. Nevertheless, patient research and advanced science have enabled us to intelligently study and investigate, and from the evidence thus gained, to state facts and formulate opinions that may perhaps outlast criticism.

The bibliographical story of "Ink" is replete with many interesting episodes, anecdotes and poetical effusions. Its chemical history is a varied and phenomenal one. Before the nineteenth century the ink industry was confined to the few. Since then, it has developed into one of magnificent proportions. The new departure, due to the discovery and development of the "Aniline" family of fugitive colors, is noteworthy as being a step backward which may take years to retrace.

The criminal abuse of ink is not infrequent by evil- disposed persons who try by secret processes to reproduce ink phenomena on ancient and modern documents. While it is possible to make a new ink look old, the methods that must be employed, will of themselves reveal to the examiner the attempted fraud, if he but knows how to investigate.

How to accomplish this as well as to give a chronological history on the subject of inks generally, both as to their genesis, the effect of time and the elements, the determination of the constituents and the constitution of inks, their value as to lasting qualities, their removal and restoration, is the object of this work. There is also included many court cases where the matter of ink was in controversy; information respecting ancient MSS. and the implements and other accessories of ink which have from time to time been employed in the act of writing.

To make a comprehensive review of the past in its relationship to ink has been my aim. In the construction of this work recourse has been had to the so- called original sources of information. In these, the diversity of their incomplete statements about different countries and epochs has offered many obstacles. In presenting my own deductions and inferences, it is with a desire to remove any impressions as to this volume being a mere compilation. "Facts are the data of all just reasoning, and the elements of all real knowledge. It follows that he is a wise man who possesses the greatest store of facts on a given subject. A book, therefore, which assembles facts from their scattered sources, may be considered as a useful and important auxiliary to those who seek them." A prolonged and continuous intercourse for over a quarter of a century with ancient and modern MSS., with books and other literature, with laymen and chemists, with students and manufacturers, together with the information and knowledge derived from experiment and study of results may enable the author to make the subject fairly clear. Effort has been made to avoid technical words and phrases in that portion treating of the Chemistry of Inks.

This work will no doubt be variously considered. Criticism is expected, indeed it is gladly invited, for thereby may follow controversy, discussion and perhaps legislation, which will bring about results beneficial to those who are to follow after us.

CONTENTS

FORTY CENTURIES OF INK

CHAPTER I.

GENESIS OF INK.

THE ORIGIN OF INK—COMPOSITION OF THE COLORED INKS OF ANTIQUITY—ANCIENT NAMES FOR BLACK INKS—METHODS OF THEIR MANUFACTURE—THE INVENTION OF "INDIAN" INK—THE ART OF DYEING HISTORICALLY CONSIDERED—THE SYMBOLIC ESTIMATION OF COLORS—THE EMPLOYMENT OF TINCTURES AS INKS—CONSIDERATION OF THE ANTIQUITY OF ARTIFICIAL INKS AND THE BLACK INKS OF INTERMEDIATE TIMES—ORIGIN OF THE COLORED PIGMENTS OF ANTIQUITY-CITATIONS FROM HERODOTUS, PLINY AND ARBUTHNOT—PRICES CURRENT, OF ANCIENT INKS AND COLORS—WHY THE NATURAL INKS FORMERLY EMPLOYED ARE NOT STILL EXTANT—THE KIND OF INK EMPLOYED BY THE PRIESTS IN THE TIME OF MOSES—ILLUSTRATIVE HISTORY OF THE EGYPTIANS IN ITS RELATIONSHIP TO WRITING IMPLEMENTS—THE USE OF BOTH RED AND BLACK INK IN JOSEPH'S TIME—ITS OTHER HISTORY PRECEDING THE DEPARTURE OF ISRAEL FROM EGYPT—THE DISAPPEARANCE OF ALL BUT A FEW KINDS OF INK—INK TRADITIONS AND THEIR VALUE—STORY ABOUT THE ORACLES OF THE SIBYLS—HOW THE ANCIENT HISTORIANS SOUGHT TO BE MISLEADING—ILLUSTRATIVE ANECDOTE BY RICHARDSON:

THE origin of Ink belongs to an era following the invention of writing. When the development of that art had advanced beyond the age of stone inscription or clay tablet, some material for marking with the reed and the brush was necessary. It was not difficult to obtain black or colored mixtures for this purpose. With their advent, forty centuries or more ago, begins the genesis of ink.

The colored inks of antiquity included the use of a variety of dyes and pigmentary colors, typical of those employed in the ancient art of dyeing, in which the Egyptians excelled and still thought by many to be one of the lost arts. The Bible and alleged contemporary and later literature make frequent mention of black and many colors of brilliant hues.

In tracing the arts of handwriting and dyeing, some definite facts are to be predicated as to the most remote history of ink.

The Hebrew word for ink is deyo, so called from its blackness. As primitively prepared for ritualistic purposes and for a continuing period of more than two thousand years, it was a simple mixture of powdered charcoal or soot with water, to which gum was sometimes added.

The Arabian methods of making ink (alchiber) were more complex. Lampblack was first made by the burning of oil, tar or rosin, which was then commingled with gum and honey and pressed into small wafers or cakes, to which water could be added when wanted for use.

About 1200 years before the Christian era, the Chinese perfected this method and invented "Indian Ink," ostensibly for blackening the surface of raised hieroglyphics, which "was obtained from the soot produced by the smoke of pines and the oil in lamps, mixed with the isinglass (gelatin) of asses' skin, and musk to correct the odour of the oil." Du Halde cites the following, as of the time of the celebrated Emperor Wu-Wong, who flourished 1120 years before Christ:

"As the stone Me (a word signifying blackening in the Chinese language), which is used to blacken the engraved characters, can never become white; so a heart blackened by vices will always retain its blackness."

That the art of dyeing was known, valued and applied among early nations, is abundantly clear. The allusions to "purple and fine raiment," to "dyed garments," to "cloth of many colours," &c., are numerous in the Bible. In a note to the "Pictorial Bible, after an allusion to the antiquity of this art, and to the pre- eminence attached by the ancients to purple beyond every other color, it is remarked: "It is important to understand that the word purple, in ancient writings, does not denote one particular colour."

Many of the names of the dyestuffs have come down to us, some of them still in use at this time and others obsolete. They were employed sometimes as ink, and certain color values given to them, of which the more important were blue, red, yellow, green, white, black, purple, gold and silver. Some colors were estimated symbolically. White was everywhere the symbol of purity and the emblem of innocence, and, just opposite, black was held up as an emblem of affliction and calamity.

Green was the emblem of freshness, vigor and prosperity.

Blue was the symbol of revelation; it was pre-eminently the celestial color blessed among heathen nations, and among the Hebrews it was the Jehovah color, the symbol of the revered God. Hence, it was the color predominant in Mosaic ceremonies.

Purple was associated as the dress of kings, with ideas of royalty and majesty.

Crimson and scarlet, from their resemblance to blood, became symbolical of life, and also an emblem of that which was indelible or deeply ingrained.

Later, in Christian times, only five colors were recognized as fitting for theological meaning or expression: white, red, green, violet and black.

White was esteemed as being the union of all the rays of light, and is often referred to as the symbol of truth and spotless purity. Red was emblematic both of fire and love, while green from its analogy to the vegetable world, was indicative of life and hope. Violet was considered the color of penitence and sorrow. Blue was forbidden except as a color peculiarly appropriated to the Virgin Mary, while black represented universally sorrow, destruction and death.

The art of dyeing was also well understood and practiced in Persia in the most ancient periods. The modern Persians have chosen Christ as their patron, and Bischoff says at present call a dyehouse Christ's workshop, from a tradition they have that He was of that profession, which is probably founded on the old legend "that Christ being put apprentice to a dyer, His master desired him to dye some pieces of cloth of different colors; He put them all into a boiler, and when the dyer took them out he was terribly frightened on finding that each had its proper color."

This, or a similar legend, occurs in the apocryphal book entitled, "The First Gospel of the Infancy of Jesus Christ." The following is the passage:

"On a certain day also, when the Lord Jesus was playing with the boys, and running about, He passed by a dyer's shop whose name was Salem, and there were in his shop many pieces of cloth belonging to the people of that city, which they designed to dye of several colors. Then the Lord, Jesus, going into the dyer's shop, took all the cloths and threw them into the furnace. When Salem came home and saw the cloth spoiled, he began to make a great noise and to chide the Lord Jesus, saying: 'What hast Thou done, unto me, O thou son of Mary? Thou hast injured both me and my neighbors; they all desired their cloths of a proper color, but Thou hast come and spoiled them all.' The Lord Jesus replied: 'I will change the color of every cloth to what color thou desirest,' and then He presently began to take the cloths out of the furnace; and they were all dyed of those same colors which the dyer desired. And when the Jews saw this surprising miracle they praised God."

The ancients used also a number of tinctures as ink, among them a brown color, sepia, in Hebrew tekeleth. As a natural ink its origin antedates every other ink, artificial or otherwise, in the world. It is a black-brown liquor, secreted by a small gland into an oval pouch, and through a connecting duct is ejected at will by the cuttle fish which inhabits the seas of Europe, especially the Mediterranean. These fish constantly employ the contents of their "ink bags" to discolor the water, when in the presence of enemies, in order to facilitate their escape from them.

The black broth of the Spartans was composed of this product. The Egyptians sometimes used it for coloring inscriptions on stone. It is the most lasting of all natural ink substances.

So great is the antiquity of artificial ink that the name of its inventor or date of its invention are alike unknown. The poet Whitehead refers to it as follows:

Hard that his name it should not save,
Who first poured forth the sable wave."

The common black ink of the ancients was essentially different in composition and less liable to fade than those used at the present time. It was not a stain like ours, and when Horace wrote

"And yet as ink the fairest paper stains,
So worthless verse pollutes the fairest deeds,"

he must have had in mind the vitriolic ink of his own time.

But little information relative to black inks of the intermediate times has come down to us, and it is conveyed through questioned writings of authors who flourished about the period of the life of Jesus Christ; the Younger Pliny and Dioscorides are the most prominent of them. They present many curious recipes. One of these, suggested by Pliny, is that the addition of an infusion of wormwood to ink will prevent the destruction of MSS. by mice.

From a memoir by M. Rousset upon the pigments and dyes used by the ancients, it would appear that the variety was very considerable. Among the white colors, they were acquainted with white lead; and for the blacks, various kinds of charcoal and soot were used. Animal skins were dyed black with gall apples and sulphate of iron (copper). Brown pigments were made by mixing different kinds of ochre. Under the name of Alexander blue, the ancients—Egyptians as well as Greeks and Romans—used a pigment containing oxide of copper, and also one containing cobalt.

Fabrics were dyed blue by means of pastel-wood.

Yellow pigments were principally derived from weld, saffron, and other native plants.

Vermilion, red ochre, and minium (red lead) were known from a remote antiquity, although the artificial preparation of vermilion was a secret possessed only by the Chinese.

The term scarlet as employed in the Old Testament was used to designate the blood-red color procured from an insect somewhat resembling cochineal, found in great quantities in Armenia and other eastern countries. The Arabian name of the insect is Kermez (whence crimson). It frequents the boughs of a species of the ilex tree: on these it lays its eggs in groups, which become covered with a sort of down, so that they present the appearance of vegetable galls or excrescences from the tree itself and are described as such by Pliny XVI, 12, who also gave it the name of granum, probably on account of its resemblance to a grain or berry, which has been adopted by more recent writers and is the origin of the term "in-grain color" as now in use. The dye is procured from the female grub alone, which, when alive is about the size of the kernel of a cherry and of a dark red-brown color, but when dead, shrivels up to the size of a grain of wheat and is covered with a bluish mold. It has an agreeable aromatic smell which it imparts to that with which it comes into contact. It was first found in general use in Europe in the tenth century. About 1550, cochineal, introduced there from Mexico, was found to be far richer in coloring matter and therefore gradually superseded the older dyestuff.

Indigo was used in India and Egypt long before the Christian era; and it is asserted that blue ribbons (strips) found on Egyptian mummies 4500 years old had been dyed with indigo. It was introduced into Europe only in the sixteenth century.

The use of madder as a red dyestuff dates from very early times. Pliny mentions it as being employed by the Hindoos, Persians and Egyptians. In the middle ages the names sandis, warantia, granza, garancia, were applied to madder, the latter (garance) being still retained in France. The color yielding substance resides almost entirely in the roots.

Chilzon was the name given by the ancient Hebrews to a blue dye procured from a species of shell-fish.

Herodotus, B. C. 443, asserts that on the shores of the Caspian Sea lived a people who painted the forms of animals on their garments with vegetable dyes:

"They have trees whose leaves possess a peculiar property; they reduce them to powder, and then strip them in water; this forms a dye or coloring matter with which they paint on their garments the figures of animals. The impression is such that it cannot be washed out; it appears, indeed, to be woven into the cloth, and wears as long as the garment itself."

We are informed by another ancient writer that the pagan nations were accustomed to array the images of their gods in robes of purple. When the prophet Ezekiel took up a lamentation for Tyre, he spoke of the "blue and purple from the isles of Elishah" in which the people were clothed. This reference is said to doubtless refer to the islands of the Aegian Sea, from whence many claim , the Tyrians obtained the shell-fish,—the murex and papura, which produced the dark-blue and bright-scarlet coloring materials, the employment of which contributed so much to the fame of ancient Tyre.

Pliny the younger confirms this statement:

"The Tyrian-purple was the juice of the Purpurea, a shell-fish, the veins of its neck and jaws secreting this royal color, but so little was obtained that it was very rare and cost one thousand Denarii (about $150.00) per pound."

A more modern writer in discussing a crimson or ruby color says:

"By a mistaken sense the Latin word purpurus, has been called purple, by all the English and French writers."

Arbuthnot, London, 1727, in his book "Ancient Coins, Weights and Measures," as the result of his examinations of the most ancient records estimates:

"The Purple was very dear; there were two sorts of Fishes whereof it was made, the Pelagii, (which were those that were caught in the deep) and the Buccini. The Pelagium per Pound was worth 50 Nummi, (8 s. 10 3/4 d.), and the Buceinunt double that, viz. 17 s. 8

3/4 d. (Harduin reads a hundred Pounds at that price.) The Tyrian double Dye per Pound could scarce be bought for L35 9 s., 1 3/4 d."

The very ancient writers state that the most esteemed of the Tyrian purples were those which compared in color with "coagulated bullocks' blood." This estimation seems to go back to the time of the Phoenicians, who were excessively fond of the redder shades of purple which they obtained also from several varieties of shell-fish and comprehended under two species; one (Buccinum) found in cliffs, and the other (Pelagia) which was captured at sea. The first was found on the coasts of the Mediterranean and Atlantic. The Atlantic shells afforded the darkest color, while those of the Phoenician coast itself yielded scarlet shades of wonderful intensity.

Respecting the cost and durability of the Tyrian purple, it is related that Alexander the Great found in the treasury of the Persian monarch 5,000 quintals of Hermione purple of great beauty, and 180 years old, and that it was worth $125 of our money per pound weight. The price of dyeing a pound of wool in the time of Augustus is given by Pliny, and that price is equal to about $160 of our money. It is probable that his remarks refer to some particular tint or quality of color easily distinguished, although not at all clearly defined by Pliny. He also mentions a sort of purple, or hyacinth, which was worth, in the time of Julius Caesar, 100 denarii (about $15 of our money) per pound.

The best authorities of the present day, however, are of opinion that the celebrated Tyrian-purple was extracted from a mollusk known as the Janthina prolongata, a shell abundant in the Mediterranean and very common near Narbonne, where the Tyrian purple dye-works were in operation at least six hundred years before Christ.

The price current of some of the inks and colors of antiquity, as quoted by Arbuthnot, are cited herewith:

Armenian purple 30 hs.=4 s. 10 1/3 d.

India purple from one Denarius, or 7 3/4 d. to 30
Denarii, 19 s. 4 1 2 d.

Pelagium, the juice of one sort fishes that dyed purple, 50 hs.=8 s. 0 7/8 d.

Buccinum the juice of the other fish that dyed purple, 100 hs.=16 s. 1 3/4 d.

Cinnabar 50 hs.=8 s. 0 7/8 d.

Tarentine red purple, price not mentioned.

Melinum, a sort of colour that came from Melos, one Nummus,=1 15/16 d.

Paretonium, a sort of colour that came from aegypt, very lasting, 6 Denarii,=3 s. 10 1/2 d.

Myrobalanus, 2 Denarii,=1 s. 3 1/2 d.

The last-named substance is the fruit of the Termi- nalia, a product of China and the East Indies, best known as Myrabolams and must have been utilized solely for the tannin they contain, which Loewe estimates to be identical with ellago-tannic acid, later discovered in the divi-divi, a fruit grown in South America, and bablah which is also a fruit of a species of Acacia, well known also for its gum.

No monuments are extant of the ancient Myrabolam ink.

Antimony and galls were used by the Egyptian ladies to tint their eyes and lashes and (who knows) to write with.

Many of the dyes employed as ink were those occurring naturally as animal and vegetable products, or which could be produced therefrom by comparatively simple means, otherwise we would not be confronted with the fact that no specimens of ink writing of natural origin remain to us.

The very few specimens of ink writing which have outlasted decay and disintegration through so many ages, are found to be closely allied to materials like bitumen, lampblack obtained from the smoke of oil- torches or resins; or gold, silver, cinnabar and minium.

Josephus asserts that the books of the ancient Hebrews were written in gold and silver.

"Sicca dewat" (A silver ink standeth), as the ancient
Arabic proverb runs.

Rosselini asserts:

"the monumental hireoglyphics of the Egyptians were almost invariably painted with the liveliest tints; and when similar hireoglyphics were executed on a reduced scale, and in a more cursive form upon papyri or scrolls made from the leaves of the papyrus the pages were written with both black and colored inks."

The early mode of ink writing in biblical times mentioned in Numbers v. 23, where It is said "the priest shall write the curses in a book, and blot them out with the bitter water," was with a kind of ink prepared for the purpose, without any salts of iron or other material which could make a permanent dye; these maledictions were then washed into the water, which the woman was obliged to drink, so that she drank the very words of the execration. The ink still used in the East is almost all of this kind; a wet sponge will obliterate the finest of their writings.

In the book of Jeremiah, chap. xxxvi. verse 18, it says: "Then Baruch answered, He pronounced all these words unto me with his mouth, and I wrote THEM with ink in the book," and in Ezek. ix. 2, 3, 11, "Ink horn" is referred to.

Six hundred years later in the New Testament is another mention of ink "having many things to write unto you. I would not write with paper and Ink," &c.; second epistle. of John, 12, and again in his third epistle, 13, "I had many things to write, but I will not with pen and Ink write unto thee."

The illustrative history of the ancient Egyptians does not point to a time before the reed was used as a pen. The various sculptures, carvings, pottery and paintings, exhibit the scribes at work in their avocations, recording details about the hands and ears of slaughtered enemies, the numbers of captives, the baskets of wheat, the numerous animals, the tribute, the treaties and the public records. These ancient scribes employed a cylindrical box for ink, with writing tablets, which were square sections of wood with lateral grooves to hold the small reeds for writing.

During the time Joseph was Viceroy of Egypt under Sethosis I, the first of the Pharaohs, B. C. 1717, he employed a small army of clerks and storekeepers throughout Egypt in his extensive grain operations. The scribes whose duties pertained to making records respecting this business, used both red and black inks, contained in different receptacles in a desk, which, when not in use, was placed in a box or trunk, with leather handles at the sides, and in this way was carried from place to place. As the scribe had two colors of ink, he needed two pens (reeds) and we see him on the monuments of Thebes, busy with one pen at work, and the other placed in that most ancient pen-rack, behind the ear. Such, says Mr. Knight, is presented in a painting at Beni Hassan.

The Historical Society of New York possesses a small bundle of these pens, with the stains of the ink yet upon them, besides a bronze knife used for making such pens (reeds), and which are alleged to belong to a period not far removed from Joseph's time. The other history of ink, long preceding the departure of Israel from Egypt, and with few exceptions until after the middle ages, can only be considered, as it is intimately bound up in the chronology and story of handwriting and writing materials. Even then it must not be supposed that the history of ink is authentic and continuous from the moment handwriting was applied to the recording of events; for the earliest records are lost to us in almost every instance. We are therefore dependent upon later writers, who made their records in the inks of their own time, and who could refer to those preceding them only by the aid of legends and traditions.

There is no independent data indicating any variation whatever in the methods of the admixture of black or colored inks, which differentiates them from those used in the earliest times of the ancient Egyptians, Hebrews or Chinese. On the contrary if we exclude "Indian" and one of the red inks, for a period of fourteen hundred years we find their number diminishing until the first centuries of the Christian era. Exaggerated tradition has described inks as well as other things and imagination is not lacking. Some of these legends, in later years put in writing, compel us to depend on translations of obscure and obsolete tongues, while the majority of them are mingled with the errors and superstitious of the time in which they were transcribed.

The value of such accounts depends upon a variety of circumstances and we must proceed with the utmost caution and discrimination in examining and weighing the authenticity of these sources of information.

If we reason that the art of handwriting did not become known to all the ancient nations at once, but was gradually imparted by one to another, it follows that records supposed to be contemporaneous, were made in some countries at a much earlier period than in others. It must also be observed that the Asiatic nations and the Egyptians practiced the art of writing many centuries before it was introduced into Europe. Hence we are able to estimate with some degree of certainty that ink-written accounts of some Asiatic nations were made while Europe was in this respect buried in utter darkness.

An interesting story which bears on this statement is told by Kennett, in his "Antiquities of Rome," London, 1743, as to the discovery of ancient MSS., five hundred and twenty years before the Christian era, of what even then must have been remarkable:

"A strange old woman came once to Tarquinius Superbus with nine books, which, she said, were the oracles of the Sybils, and proffered to sell them. But the king making some scruple about the price, she went away and burnt three of them; and returning with the six, asked the same sum as before. Tarquin only laughed at the humour; upon which the old woman left him once more; and after she had burnt three others, came again with them that were left, but still kept to her old terms. The king now began to wonder at her obstinacy, and thinking there might be something more than ordinary in the business, sent for the augars (soothsayers) to consult what was to be done. They, when their divinations were performed, soon acquainted him what a piece of impiety he had been guilty of, by refusing a treasure sent to him from heaven, and commanded him to give whatever she demanded for the books that remained. The woman received her money, and delivered the writings; and only, charging them by all means to keep them sacred, immediately vanished. Two of the nobility were presently after chosen to be the keepers of these oracles, which were laid up with all imaginable care in the Capitol, in a chest under ground. They could not be consulted without a special order of the Senate, which was never granted, unless upon the receiving of some notable defeat; upon the rising of any considerable mutiny, or sedition in the State; or upon some other extraordinary occasion; several of which we meet with in Livy."

Some of the ancient historians even sought to be misleading respecting the events not only of their own times, but of epochs which preceded them. Richardson, in his "Dissertation on Ancient History and Mythology," published in 1778, remarks:

"The information received hitherto has been almost entirely derived through the medium of the Grecian writers; whose elegance of taste, harmony of language, and fine arrangement of ideas, have captivated the imagination, misled the judgment, and stamped with the dignified title of history, the amusing excursions of fanciful romance. Too proud to consider surrounding nations, (if the Eyptians may be excepted) in any light but that of barbarians; they despised their records, they altered their language, and framed too often their details, more to the prejudices of their fellow citizens, than to the standard of truth or probability. We have names of Persian kings, which a Persian could not pronounce; we have facts related they apparently never knew; and we have customs ascribed to them, which contradict every distinguishing characteristic of an Eastern people. The story of Lysimachus and one Greek historian may indeed, with justice, be applied to many others. This prince, in the partition of Alexander's empire, became King of Thrace: he had been one of the most active of that conqueror's commanders; and was present at every event which deserved the attention of history. A Grecian had written an account of the Persian conquest; and be wished to read it before the king. The monarch listened with equal attention and wonder: 'All this is very fine,' says he, when the historian had finished, 'but where was I when those things were performed?' "

CHAPTER II.

ANTIQUITY OF INK.

THE INVENTION OF THE ART OF WRITING—TO WHOM IT BELONGS—ITS UTILIZATION BY NATIONS AND INDIVIDUALS—WHEN IT IS FIRST MENTIONED IN THE BIBLE—CITATIONS FROM THE ENCYCLOPaeDIA BRITANNICA AND SMITHS DICTIONARY OF THE BIBLE—SOME REMARKS BY HUMPHREYS OF THE ORIGIN AND PROGRESS OF HANDWRITING—COMMENTS BY PLATO AND THE COLLOQUY BETWEEN KING THAMUS AND THOTH, THE EGYPTIAN GOD OF THE LIBERAL ARTS—FIRST APPEARANCE OF INK WRITTEN ROLLS—DESTRUCTION OF THE TEMPLES WHICH CONTAINED THEM—COMMENTS OF THE HISTORIAN ROLLINS—DESTRUCTION OF THE MOST ANCIENT CHINESE INK WRITTEN MSS.

THERE is a difference of opinion as to what nation belongs the honor of the invention of the art of handwriting. Sir Isaac Newton observes:

"There is the utmost uncertainty in the chronology of ancient kingdoms, arising from the vanity of each claiming the greatest antiquity, while those pretensions were favoured by their having no exact account of time."

Its antiquity has been exhaustively treated by many writers; the best known are Massey, 1763, The Origin and Progress of Letters;" Astle, 1803, "The Origin and Progress of Writing;" Silvestre, "Universal Palaeography," Paris, 1839-41 ; and Humphreys, 1855, "The Origin and Progress of the Art of Writing." They, with others, have sought to record the origin and gradual development of the art of writing from the Egyptian Hieroglyphics of 4000 B. C.; the Chinese Figurative, 3000 B. C. ; Indian Alphabetic, 2000 or more B. C. ; the Babylonian or Cuneiform, 2000 years B. C.; and the Phoenician in which they include the Hebrew or Samaritan Alphabet, 2000 or more B. C., down to the writings of the new or Western world of the Christian era.

The data presented and the arguments set forth, deserve profound respect, and though we find some favoring the Egyptians, or the Phoenicians, the Chaldeans, the Syrians, the Indians, the Persians or the Arabians, it is best to accept the concensus of their opinion, which seems to divide between the Phoenicians and the Egyptians as being the inventors of the foremost of all the arts. "For, in Phoenicia, had lived Taaut or Thoth the first Hermes, its inventor, and who later carried his art into Egypt where they first wrote in pictures, some 2200 years B. C."

The art appears to have been first exercised in Greece and the West about 1500 or 1800 B. C., and like all arts, it was doubtless slow and progressive. The Greeks refer the invention of written letters to Cadmus, merely because he introduced them from Phoenicia, then only sixteen in number. To these, four more were added by Simonides. Evander brought letters into Latium from Greece, the Latin letters being at first nearly the same form as the Greek. The Romans employed a device of scattering green sand upon tables, for the teaching of arithmetic and writing, and in India a "sand box" consisting of a surface of sand laid on a board the finger being utilized to trace forms, was the method followed by the natives to teach their children. It is said that such methods still obtain even in this age, in some rural districts of England.

After the invention of writing well-informed nations and individuals kept scribes or chroniclers to record in writing, historical and other events, mingled with claims of antiquity based on popular legends.

These individuals were not always held in the highest esteem. Among the Hebrews it was considered an honorable vocation, while the Greeks for a long time treated its practitioners as outcasts. It was an accomplishment possessed by the few even down to the fifteenth century of the Christian era. The rulers of the different countries were deficient in the art and depended on others to write their documents and letters to which they appended their monogram or the sign of the Cross against their names as an attestation. So late as A. D. 1516 an order was made in London to examine all persons who could write in order to discover the authorship of a seditious document.

The art of writing is not mentioned in the Bible prior to the time of Moses, although as before stated, in Egypt and the countries adjacent thereto it was not only known but practiced.

Its first mention recorded in Scripture will be found in Exodus xvii. v. 14; "And the Lord said unto Moses, Write this, for a memorial, in a book; and rehearse it in the ear of Joshua; for I will utterly put out the remembrance of Amalek from under heaven." This command was given immediately after the defeat of the Amalekites near Horeb, and before the arrival of the Israelites at Mount Sinai.

It is observable, that there is not the least hint to induce us to believe that writing was then newly invented; on the contrary, we may conclude, that Moses understood what was meant by writing in a book; otherwise God would have instructed him, as he had done Noah in building the Ark; for he would not have been commanded to write in a book, if he had been ignorant of the art of writing; but Moses expressed no difficulty of comprehension when he received this command. We also find that Moses wrote all the works and all the judgments of the Lord, contained in the twenty-first and the two succeeding chapters of the book of Exodus, before the two written tables of stone were even so much as promised. The delivery of the tables is not mentioned till the eighteenth verse of the thirty-first chapter, after God had made an end of communing with him upon the mount, though the ten commandments were promulgated immediately after his third descent.

Moses makes frequent mention of ancient books of the Hebrews, but describes none, except the two tables on which God wrote the ten commandments. These he tells us, were of polished stone, engraven on both sides and as Calmet remarks: "it is probable that Moses would not have observed to us these two particulars so often as he does, were it not to distinguish them from other books, which were made of tables, not of stone, but of wood and curiously engraven, but on one side only."

It cannot be said that Moses uses any language which can be construed to mean the employment of rolls of papyrus, or barks of trees, much less of parchment. We have therefore reason to believe that by the term book, he always means table-books, made of small thin boards or plates.

The edicts, as well as the letters of kings, were written upon tablets and sent to the various provinces, sealed with their signets. Scripture plainly alludes to the custom of sealing up letters, edicts and the tablets on which the prophets wrote their visions.

The practice of writing upon rolls made of the barks of trees is very ancient. It is alluded to in the Book of Job: "Oh! that mine adversary had written a book; surely I would take it upon my shoulders, and bind it as a crown to me." (Old version.) The new one runs: "And that I had the indictment which mine adversary hath written!" The rolls, or volumes, generally speaking, were written upon one side only. This is intimated by Ezekiel who observes that he saw one of in extraordinary form written on both sides: "And when I looked, behold, an Hand was sent unto me, and lo! a roll of a book was therein; and he spread it before me, and it was written within and without."

To have been able to write on dry tablets of wood or barks of trees with the reed or brush, the then only ink-writing instruments in vogue would have necessitated the employment of lampblack suspended in a vehicle of thick gum, or in the form of a paint. Both of these maybe termed pigmentary inks. The use of thin inks would have caused spreading or blotting and thus rendered the writing illegible.

The Encyclopaedia Britannica generalizes its remarks on this subject:—

"The earliest writings were purely monumental and accordingly those materials were chosen which were supposed to last the longest. The same idea of perpetuity which in architecture finds its most striking exposition in the pyramids was repeated, in the case of literary records, in the two columns mentioned by Josephus, the one of stone and the other of brick, on which the children of Seth wrote their inventions and astronomical discoveries; in the pillars in Crete on which, according to Porphyry, the ceremonies of the Corybantes were inscribed; in the leaden tablets containinlu the works of Hesiod, deposited in the temple of the Muses, in Boeotia; in the ten commandments on stone delivered by Moses; and in the laws of Solon, inscribed on planks of wood. The notion of a literary production surviving the destruction of the materials on which it was first written—the 'momentum, aere perennius' of Horace's ambition—was unknown before the discovery of substances for systematic transcription.

"Tablets of ivory or metal were in common use among the Greeks and Romans. When made of wood—sometimes of citron, but usually of beech or fir—their inner sides were coated with wax, on which the letters were traced with a pointed pen or stiletto (stylus), one end of which was used for erasure. It was with his stylus that Caesar stabbed Casca in the arm when attacked by his murderers. Wax tablets of this kind continued in partial use in Europe during the middle ages; the oldest extant specimen, now in the museum at Florence, belongs to the year 1301."

Later the Hebrew Scriptures were written in ink or paint upon the skins of ceremonially clean animals or even birds. These were rolled upon sticks and fastened with a cord, the ends of which were sealed when security was an object. They were written in columns, and usually upon one side, only. The writing was from right to left; the upper margin was three fingers broad, the lower one four fingers; a breadth of two fingers separated the columns. The columns ran across the width of the sheet, the rolled ends of which were held vertically in the respective hands. When one column was read, another was exposed to view by unrolling it from the end in the left hand, while the former was hidden from view by rolling up the end grasped by the right band. The pen was a reed, the ink black, carried in a bottle suspended from the girdle.

The Samaritan Pentateuch is very ancient, as is proved by the criticisms of Talmudic writers. A copy of it was acquired in 1616 by Pietro della Valle, one of the first discoverers of the cuneiform inscriptions. It was thus introduced to the notice of Europe. It is claimed by the Samaritans of Nablus that their copy was written by Abisha, the great-grandson of Aaron, in the thirteenth year of the settlement of the land of Canaan by the children of Israel. The copies of it brought to Europe are all written in black ink on vellum or "cotton" paper, and vary from 12mo to folio. The scroll used by the Samaritans is written in gold letters. (See Smith's "Dictionary of the Bible," vol. III, pp. 1106-1118.) Its claims to great antiquity are not admitted by scholars.

The enumeration of some of the modes of writing may be interesting:

The Mexican writing is in vertical columns, beginning at the bottom.

The Chinese and Japanese write in vertical columns, beginning at the top and passing from left to right.

The Egyptian hieroglyphics are written invertical columns or horizontal lines according to the shape and position of the tablet. It is said that with the horizontal writing the direction is indifferent, but that the figures of men and animals face the beginning of the line. With figures, the units stand on the left.

The Egyptians also wrote from right to left in the hieratic and demotic and enchorial styles. The Palasgians did the same, and were followed by the Etruscans. In the demotic character, Dr. Brugsch remarks that though the general direction of the writing was usually from right to left, yet the individual letters were formed from left to right, as is evident from the unfinished ends of horizontal letters when the ink failed in the pen.

In writing numbers in the hieratic and enchorial the units were placed to the left. The Arabs write from right to left, but received their numerals from India, whence they call them "Hindee," and there the arrangement of their numerals is like our own, units to the right.

The following noteworthy passage is taken from Humphreys' work "On the Origin and Progress of the Art of Writing:"

"Nearly all the principal methods of ancient writing may be divided into square capitals, rounded capitals, and cursive letters; the square capitals being termed simply capitals, the rounded capitals uncials, and the small letters, or such as had changed their form during the creation of a running hand, minuscule. Capitals are, strictly speaking, such letters as retain the earliest settled form of an alphabet; being generally of such angular shapes as could conveniently be carved on wood or stone, or engraved in metal, to be stamped on coins. The earliest Latin MSS. known are written entirely in capitals like inscriptions in metal or marble. * * * * *

The uncial letters, as they are termed, appear to have arisen as writing on papyrus or vellum became common, when many of the straight lines of the capitals, in that kind of writing, gradually acquired a curved form, to facilitate their more rapid execution. However this may be, from the sixth to the eighth, or even 10th century, these uncials or partly rounded capitals prevail.

"The modern minuscule, differing from the ancient cursive character, appears to have arisen in the following manner: During the 6th and 7th centuries, a kind of transition style prevailed in Italy and some other parts of Europe, the letters composing which have been termed semi-uncials, which, in a further transition, became more like those of the old Roman cursive. This manner, when definitely formed, became what is now termed the minuscule manner; it began to prevail over uncials in a certain class of MSS. about the 8th century, and towards the 10th its general use was, with few exceptions, established. It is said to have been occasionally used as early as the 5th century; but I am unable to cite an authentic existing monument. The Psalter of Alfred the Great, written in the 9th century, is in a small Roman cursive hand, which has induced Casley to consider it the work of some Italian ecclesiastic."

The learned who have made a life study of the history of the most ancient manuscripts, mention them specifically in great number and of different countries, which would seem to indicate that the art of handwriting had made great strides in the very olden times; many nations had adopted it, and B. C. 650 "it had spread itself over the (then known) greater part of the civilized world."

We can well believe this to be true in reading about the ancient libraries, notwithstanding that some rulers had sought to prohibit its exercise.

Plato, who lived B. C. 350, expresses his views of the importance of writing in his imaginary colloquy between Thamus, king of Egypt, and Thoth, the god of the liberal arts of the Egyptians; he acquaints us:

"That the discourse turned upon letters. Thoth maintained the value of Writing, as capable of making the People wiser, increasing the powers of Memory; to this the king dissented, and expressed his opinion that by the exercise of this Art the multitude would appear to be knowing of those things of which they were really ignorant, possessing only an idea of Wisdom, instead of Wisdom itself."

Pythagoras, B. C. 532, we are informed by Astle:

"Went into Egypt where he resided twenty-two years; he was initiated into the sacerdotal order, and, from his spirit of inquiry, he has been justly said to have acquired a great deal of Egyptian learning, which he afterwards introduced into Italy. The Pythagorean schools which he established in Italy when writing was taught, were destroyed when the Platonic or new philosophy prevailed over the former. Polybius (lib. ii. p. 175) and Jamblichus (in vita Pythag.) mention many circumstances, relative to these facts, quoted from authors now lost; as doth Porphyry, in his life of Pythagoras."

For the hundred years or more following, however, the dissemination of learning and the transcription of events was not to be denied. We find ink-written volumes (rolls) relating to diverse subjects being loaned to one another; correspondence by letter to and from distant lands of frequent occurrence, and the art of handwriting regularly taught in the schools of learning. Its progress was to be interrupted by the wars of the Persians. Mr. Astle in calling attention to events which have contributed to deprive us of the literary treasures of antiquity thus refers to them:

"A very fatal blow was given to literature, by the destruction of the Phoenician temples, and of the Egyptian colleges, when those kingdoms, and the countries adjacent, were conquered by the Persians, about three hundred and fifty years before Christ. Ochus, the Persian general, ravaged these countries without mercy, and forty thousand Sidonians burnt themselves with their families and riches in their own houses. The conqueror then drove Nectanebus out of Egypt, and committed the like ravages in that country; afterwards he marched into Judea, where he took Jericho, and sent a great number of Jews into captivity. The Persians had a great dislike to the religion of the Phoenicians and the Egyptians; this was one reason for destroying their books, of which Eusebius (De Preparat. Evang.) says, they had a great number."

These losses, apparently, did not interfere with the progress of the art in more western countries. Professor Rollin in his "Ancient History," 1823, remarks:

"Ptolemy Soter, King of Egypt B. C. 285, had been careful to improve himself in public literature, as was evident by his compiling the life of Alexander, which was greatly esteemed by the ancients, but is now entirely lost. In order to encourage the cultivation of the sciences, which he much admired, he founded an academy at Alexandria, called the Museum, where a society of learned men devoted themselves to philosophic studies, and the improvement of all other sciences, almost in the same manner as those of London and Paris. For this purpose, he began by giving them a library, which was prodigiously increased by his successors.

"His son Philadelphus left a hundred thousand volumes in it at the time of his death, and the succeeding princes of that race enlarged it still more, till at last it consisted of seven hundred thousand volumes.

"This library was formed by the following method: All the Greek and other books that were brought into Egypt were seized, and sent to the Museum, where they were transcribed by persons employed for that purpose. The copies were then delivered to the proprietors, and the originals were deposited in the library.

"As the Museum was at first in that quarter of the city which was called Bruchion, and near the royal palace, the library was founded in the same place, and it soon drew vast numbers thither; but when it was so much augmented, as to contain four hundred thousand volumes, they began to deposit the additional books in the Serapion. This last library was a supplement to the former, for which reason it received the appellation of its Daughter, and in process of time had in it three hundred thousand volumes.

"In Caesar's war with the inhabitants of Alexandria, a fire, occasioned by those hostilities, consumed the library of Bruchion, with its four hundred thousand volumes. Seneca seems to me to be out of humour, when, speaking of the conflagration, he bestows his censures both on the library itself, and the eulogium made on it by Livy, who styles it an illustrious monument of the opulence of the Egyptian kings, and of their judicious attention to the improvement of the sciences. Seneca, instead of allowing it to be such, would have it considered only as a work resulting from the pride and vanity of those monarchs, who had amassed such a number of books, not for their own use, but merely for pomp and ostentation. This reflection, however, seems to discover very little sagacity; for is it not evident beyond contradiction, that none but kings are capable of founding these magnificent libraries, which become a necessary treasure to the learned, and do infinite honour to those states in which they are established?

"The library of Serapion, did not sustain any damage, and it was undoubtedly there that Cleopatra deposited those two hundred thousand volumes from that of Pergamus, which was presented to her by Antony. This addition, with other enlargements that were made from time to time, rendered the new library of Alexandria more numerous and considerable than the first; and though it was ransacked more than once, during the troubles and revolutions which happened in the Roman empire, it always retrieved its losses, and recovered its number of volumes. In this condition it subsisted for many ages, displaying its treasures to the learned and curious, till the seventh century, when it suffered the same fate with its parent, and was burnt by the Saracens, when they took that city in the year of our Lord 642. The manner by which this misfortune happened is too singular to be passed over in silence.

"John, surnamed the Grammarian, a famous follower of Aristotle, happened to be at Alexandria, when the city was taken; and as he was much esteemed by Amri Ebnol As, the general of the Saracen troops, he entreated that commander to bestow upon him the Alexandrian library. Amri replied, that it was not in his power to grant such a request; but that he would write to the Khalif, or emperor of the Saracens, for his orders on that head, without which he could not presume to dispose of the library. He accordingly wrote to Omar, the then Khalif, whose answer was, that if those books contained the same doctrine with the Koran, they could not be of any use, because the Koran was sufficient in itself, and comprehended all necessary truths; but if they contained any particulars contrary to that book, they ought to be destroyed. In consequence to this answer, they were all condemned to the flames, without any further examination; and, for that purpose, were distributed among the public baths; where, for the space of six months, they were used for fuel instead of wood. We may from hence form a just idea of the prodigious number of books contained in that library; and thus was this inestimable treasure of learning destroyed!

The Museum of Bruchion was not burnt with the library which was attached to it. Strabo acquaints us, in his description of it, that it was a very large structure near the palace, and fronting the port; and that it was surrounded with a portico, in which the philosophers walked. He adds, that the members of this society were governed by a president, whose station was so honourable and important, that, in the time of the Ptolemies, he was always chosen by the king himself, and afterwards by the Roman emperor; and that they had a hall where the whole society ate together at the expense of the public, by whom they were supported in a very plentiful manner."

Among the other events contributing to the deplorable losses which mankind has sustained in this respect, a sad one was when the most ancient ink writings of the Chinese were ordered to be destroyed by their emperor Chee-Whange-Tee, in the third century before Christ, with the avowed purpose that everything should begin anew as from his reign. The small portion of them which escaped destruction were recovered and preserved by his successors.

CHAPTER III.

CLASSICAL INK AND ITS EXODUS.

THE MATERIALS AND METHODS EMPLOYED IN PREPARING THE INK MSS. OF ANTIQUITY—THE INTRODUCTION OF PARCHMENT AS A SUBSTITUTE FOR PAPYRUS—MODE OF WRITING ON PARCHMENT—HOW SEPARATE PIECES WERE FIRST JOINED INTO BOOK FORM—EVIDENCE OF THE CHARACTER OF WRITING UTENSILS TO BE FOUND IN

ANCIENT PICTURES—SOME FORMULAS BY THE YOUNGER PLINY AND HIS CONTEMPORARY DIOSCORIDES—HOW THE GREEKS AND ROMANS KEPT THEIR PAPYRI FROM BREAKING—WHEN BLACK INK BEGAN TO FALL INTO DISUSE AND ITS CAUSE—THE ADOPTION OF THE STYLUS AND ITS ACCOMPANYING SHEETS OF LEAD, IVORY, METAL AND WOOD COATED WITH WAX—THE EFFORTS MADE TO RESUME THE USE OF SOME INK WHICH WOULD BIND TO PARCHMENT—WHY THERE ARE NO ORIGINAL MSS. EXTANT BELONGING TO THE TIME OF CHRIST—THE INVENTION OF THE VITRIOLIC INKS—HUMPHREY'S BLUNDER IN LOCATING DATES OF EARLY GREEK MSS.—THE DESTRUCTION OF THE CITIES OF HERCULANEUM AND POMPEII—AWAKENING OF INTEREST AGAIN ABOUT THE EMPLOYMENT OF INKS—REDISCOVERIES OF SOME OF THE MORE REMOTE ANCIENT RECIPES—THE WRITERS IN GOLD AND SILVER—RECORDED INSTANCES OF ILLUMINATED MSS.—PASSAGE FROM THE BOOK OF JOB WRITTEN BY ST. JEROME—DENIAL OF THE EMPLOYMENT OF TANNO- GALLATE OF IRON INK IN THE FOURTH CENTURY— DESTRUCTION OF THE INSPIRED WRITINGS BY ORDER OF THE ROMAN SENATE—THE ECLIPSE OF CLASSICAL LITERATURE AND DISMEMBERMENT OF THE ROMAN EMPIRE—POEM ON THE THOUSAND YEARS KNOWN AS THE DARK AGES WHICH FOLLOWED.

THEOPHRASTUS says that the papyrus books of the ancients were no other than rolls prepared in the following manner: Two leaves of the rush were plastered together, usually with the mud of the Nile, in such a fashion that the fibres of one leaf should cross the fibres of the other at right angles; the ends of each being then cut off, a square leaf was obtained, equally capable of resisting fracture when pulled or taken hold of in any direction. In this form the papyri were exported in great quantities. In order to form these single leaves into the "scapi," or rolls of the ancients, about twenty were glued together end to end. The writing was then executed in parallel columns a few inches wide, running transversely to the length of the scroll. To each end of the scrolls were attached round staves similar to those we use for maps. To these staves, strings, known as "umbilici," were attached, to the ends of which bullae or weights were fixed. The books when rolled up, were bound up with these umbilici, and were generally kept in cylindrical boxes or capsae, a term from which the Mediaeval "capsula," or book-cover was derived. "The mode in which the students held the rolls in order to read from them is well shown in a painting in the house of a surgeon at Pompeii. One of the staves, with the papyrus rolled round it, was held in each hand, at a distance apart equal to the width of one or more of the transverse columns of writing. As soon as the eye was carried down to the bottom of a column, one hand rolled up and the other unrolled sufficient of the papyrus to bring a fresh column opposite to the reader's eye, and so on until the whole was wound round one of the staves, when, of course, the student had arrived at the end of his book."

Eumenes, king of Pergamus, being unable to procure the Egyptian papyrus, through the jealousy of one of the Ptolemies, who occupied himself in forming a rival library to the one which subsequently became so celebrated at Pergamus, introduced the use of Parchment properly "dressed" for taking ink and pigments and hence the derivation of the word "pergamena" as applied to parchment or vellum, the former substance being the prepared skin of sheep, and the latter of calves.

The sheets of parchment were joined end to end, as the sheets of papyrus had been, and when written upon, on one side only, and in narrow columns across the breadth of the scroll, were rolled up around staves and bound with strings, to which seals of wax were occasionally attached, in place of the more common leaden bullae.

The custom of dividing wax, ivory, wood and metal MSS. into pages and in this way into book form is said by Suetonius to have been introduced by Julius Caesar, whose letters to the Senate were so made up, and after whose time the practice became usual for all documents either addressed to, or issuing from that body, or to or from the Emperors. As that form subsequently crept into general use, the books were known as "codices;" and hence the ordinary term as applied to manuscript volumes.

All classes of "books," the reeds for writing in them, the inkstands, and the "capsae" or "scrinia," the boxes in which the "scapi" or rolls were kept, are minutely portrayed in ancient wall-paintings and ivory diptychs (double tablets), and which may belong to a period near the beginning of the Christian era.

Pliny and Dioscorides have given the formulas for the writing inks used by the Greek and Roman scribes immediately before and during their time. Pliny declares that the ink of the bookmakers was made of soot, charcoal and gum, although he does not state what fluid was employed to commingle them. He does, however, mention to an occasional use of some acid (vinegar) to give the ink a binding property on the papyrus.

Dioscorides, however, specifies the proportions of this "soot" ink. Another formula alluded to by the same author calls for a half ounce each of copperas (blue) and ox-glue, with half pound of smoke black made from burned resin. He adds, "is a good application in cases of gangrene and is useful in scalds, if a little thickened and employed as a salve." De Vinne speaks of this as a "crude" receipt which will enable one to form a correct opinion of the quality of scientific knowledge then applied to medicine and the mechanical arts; also that these mixtures which are more like shoe blacking than writing fluid were used with immaterial modifications by the scribes of the dark ages.

The old Greeks and Romans had no substitute for the papyrus, which was so brittle that it could not be folded or creased. It could not be bound up in books, nor could it be rolled up unsupported. It was secure only when it had been wound around a wooden or metal roller.

After the wholesale destruction of the libraries of ink-written MSS., the black inks began to fall into disuse; their value in respect to quality gradually deteriorated, caused by the displacement of gummy vehicles, and a consequent absence of any chance of union between the parchment or papyrus and the dry black particles, which could be "blown" or washed off. To employ any other kind of ink except one of natural origin like the juice of berries which soon disappeared, was forbidden by prevailing religious customs. Such conditions naturally merged into others, in the shape of "ink" substitutes for writing; the stylus, with its accompanying sheets or tablets of ivory, wood, metal and wax came into popular vogue and so continued for many centuries, even after the employment of ink for writing purposes had been resumed.

Ovid, in his story of Caunus and Byblis, illustrates the use of the tables (tablets), and he lived at the time of the birth of Christ, thus translated:

"Then fits her trembling hands to Write:
One holds the Wax, the Style the other guides,
Begins, doubts, writes, and at the Table chides;
Notes, razes, changes oft, dislikes, approves,
Throws all aside, resumes what she removes.
* * * * * * * *
"The Wax thus filled with her successless wit,
She Verses in the utmost margin writ."

He also makes reference to inks, in the passage taken from his first elegy, "Ad Librum:"

"Nec te purpureo velent vaccinia succo;
Non est conveniens luctibus ille color.
Nec titulus minio, nec cedro charta notetur.
Candida nec nigra cornua fronte geras."

which Davids translates as follows:

"TO HIS BOOK.

"Nor shall huckleberries stain (literally veil) thee with purple juice:
That color is not becoming to lamentations.
Nor shall title (or head-letter) be marked with vermillion, or paper with cedar,
Thou shalt carry neither white nor black horns on thy forehead (or front, or frontispiece)."

The traditions handed down as of this era relating to the efforts to find some substitute for "Indian" ink which would not only "bind" to parchment and vellum but also would be satisfactory to the priests, are more or less confirmed by the younger Pliny, and makes it safe to assume that several were invented and employed in writing, though possessing but little lasting qualities. Their use and natural disappearance is perhaps the real cause of the fact that there are no original MSS. extant dating as of or belonging to the time immediately preceding or following the birth of Christ, or indeed until long after his death.

There is some authority though for the statement that at this time two vitriolic substances were used in the preparation of black ink,—a slime or sediment (Salsugo) and a yellow vitriolic earth (Misy). This last-named mineral, is unquestionably the same natural chemical mentioned by writers, which about the end of the first century was designated "kalkanthum" or "chalkanthum" and possessed not only the appearance of, but the virtues of what we know as blue copperas or sulphate of copper. It continued in use as long as men were unacquainted with the art of lixiviating salt, or, in other words, as long as they had no vitriol manufactories. Commingled with lampblack, bitumen or like black substances in gummy water, it was acceptable to the priests for ritualistic writings and was in general vogue for several centuries thereafter under the name of (blue) "vitriolic" ink, notwithstanding the fact that there could not be any lasting chemical union between such materials.

It was the so-called "vitriolic" ink, which is said to have "corroded the delicate leaves of the papyrus and to have eaten through both parchment and vellum."

These deductions, however, do not agree with some of the historians and scholars like Noel Humphreys, author of the "Origin and Progress of the Art of Writing," London, 1855, a recognized authority on the subject of ancient MSS., who but repeats in part the text of earlier writers, when he says, p. 101:

"Examples of early Greek MSS. of the last century previous to the Christian era are not confined to Egyptian sources; the buried city of Herculaneum, in Italy, partially destroyed about seventy- nine years before the Christian era, and injured by subsequeut eruptions, till totally destroyed by the most violent eruption of Vesuvius on record, that of the year 471 A. D. having yielded several specimens."

The MSS. examples mentioned in the citation, must of necessity refer to specimens of writing made with "vitriolic" and even more ancient inks. They are to be considered in conjunction with the historical fact that these cities were buried for more than sixteen hundred years, counting from the first eruption, before they were brought to light (Herculaneum was discovered A. D. 1713 and Pompeii, forty years later); also that they must have been subjected to intense heat and a long period of decay which could only operate to rob them of all traces of natural ink phenomena. Furthermore, the information Mr. Humphreys seeks to convey, dates contemporaneously with the first eruption of Vesuvius, which occurred seventy-nine years AFTER the Christian era and not seventy-nine years BEFORE it.

This stupendous blunder involves a period of one hundred and fifty-eight years; if it is rectified, the "early Greek MSS." are shown to emanate from the second half of the first century following the birth of Christ and confirming to some extent the deductions hereinbefore made, although the probabilities are that they belong to later periods, included in the third and fourth centuries.

It is affirmed that the eruption of Mt. Vesuvius A. D. 79, did not entirely destroy the cities of Herculaneum and Pompeii, and that they emerged from their ruins in the reign of the Emperor Titus. They are also mentioned as inhabited cities in the chart of Peutinger, which is of the date of Constantine.

The next eruption, A. D. 471, was probably the most frightful on record if we exclude the volcanic eruption of Mt. Pelee, which occurred in Martinique, West Indies, in 1902, destroying thirty thousand human beings in fifteen minutes and devastating nearly the entire island. From Marcellinus we learn that the ashes of the Vesuvius volcano were vomited over a great portion of Europe, reaching to Constantinople, where a festival was instituted in commemoration of the strange phenomenon. After this, we hear no more of these cities, but the portion of the inhabitants who escaped built or occupied suburbs at Nola in Campania and at Naples. In the latter city, the Regio Herculanensium, or Quarter of the Herculaneans, an inscription marked on several lapidary monuments, indicates the part devoted to the population driven from the doomed city.

The ancient inkstand found at Herculaneum, said to contain a substance resembling a thick oil or paint characteristic of a material which it is alleged, "some of the manuscripts have been written in a sort of relievo, visible in the letters when a 'leaf' is held to the light in a horizontal direction," it is not impossible, indeed it is quite probable, belonged to an era centuries later than the period to which it has been assigned.

"No perfect papyri, but only fragments, have been found at Pompeii. At Herculaneum, up to the year 1825, 1,756 had been obtained, besides many others destroyed by the workmen, who imagined them to be mere sticks of charcoal. Most of them were found in a suburban villa, in a room of small dimensions, ranged in presses round the sides of the room, in the center of which stood a sort of rectangular bookcase.

"Sir Humphry Davy, after investigating their chemical nature, arrived at the conclusion that they had not been carbonized by heat, but changed by the long action of air and moisture; and he visited Naples in hopes of rendering the resources of chemistry available towards deciphering these long-lost literary treasures. His expectations, however, were not fully crowned with success, although the partial efficacy of his methods was established; and he relinquished the pursuit at the end of six months, partly from disappointment, partly from a belief that vexatious obstacles were thrown in his way by the jealousy of the persons to whom the task of unrolling had been intrusted. About five hundred volumes have been well and neatly unrolled. It is rather remarkable that, as far as can be learned, no manuscript of any known standard work has been found, nor, indeed, any production of any of the great luminaries of the ancient world. The most celebrated person of whom any work has been found is Epicurus, whose treatise, De Natura, has been successfully unrolled. This and a few other treatises have been published. The library in which this was found appears to have been rich in treatises on the Epicurean philosophy. The only Latin work which it contained was a poem, attributed to Rabirius, on the war of Caesar and Antony."

Beginning with A. D. 200, the employment of inks became more and more constant and popular. Rediscoveries of ancient formulas belonging to a more remote antiquity multiplied in number. Silver ink was again quite common in most countries. Red ink made of vermilion (a composition of mercury, sulphur and potash) and cinnabar (native mercuric sulphide) were employed in the writing of the titles as was blue ink made of indigo, cobalt or oxide of copper. Tyrian purple was used for coloring the parchment or vellum. The "Indian" inks made by the Chinese were imported and used in preference to those of similar character manufactured at home. The stylus and waxed tablets though still used, in a measure gave way to the reawakened interest in ink and ink writings.

A greater facility in writing, due to the gradual reduction in size of the uncial (inch) letters was thereby attained.

There were "writers in gold" and "writers in silver" who travelled from the East into Greece and who bad found their way before the third century into the very heart of Rome. Their business was to embellish the manuscript writings of those times. It was considered en regale for authors to "illuminate" their MSS. and those who failed to do so suffered in popularity.

These authors frequently allude to their use of red, black and secret inks.

Martial in his first epistle points out the bookseller's shop opposite the Julian Forum where his works may be obtained "smoothed with pumice stone and decorated with purple." Seneca mentions books ornamented "cum imaginabus." Varro is related by the younger Pliny to have illustrated his works by pictures of more than seven hundred illustrious persons. Martial dwells on the edition of Virgil, with his portrait as a frontispiece.

The earliest recorded instance of the richer adornments of golden lettering on purple or rose-stained vellum is given by Julius Capitolinus in his life of the Emperor Maximinus the younger. He therein mentions that the mother of the emperor presented to him on his return to his tutor (early in the third century), a copy of the works of Homer, written in gold upon purple vellum.

The fugitive character, as before stated, of a great many of the colored inks, and indeed most of the black ones which were undoubtedly employed, is the principal reason why so few specimens of them remain to us. Those which have proved themselves so lasting in character as to be still extant, bear evidence of extreme care in the preparation of both the inks and the materials on which the writings appear. Perhaps one of the finest illustrations of this practice is to be found in a book of the Four Gospels of Italian origin, discovered in the tenth century (a work of the fourth century) and deposited in the Harlein Library. This book is written in "Indian" ink and possesses magnificently embellished and illuminated letters at the beginning of each Gospel, which are on vellum stained in different colors.

St. Jerome calls attention to this class of books in a well-known passage of his preface to the Book of Job, also written in the fourth century, where he explains as translated:

"Let those who will have old books written in gold and silver on purple parchment, or, as they are commonly called, in uncial-letters,—rather ponderous loads than books,—so long as they permit me and mine to have copies, and rather correct than beautiful books."

It has been said that the Tanno-gallate of Iron Inks (iron salts, nutgalls and gum) were first used in the fourth century. There is positively no credible authority for such a statement, nor is there a single monument in the shape of a documentary specimen of ink writing of that one or an earlier century made with such an ink in any public or private library and as far as known in existence.

About A. D. 390 the inspired writings (often termed pagan) of the classical countries, or at least the copies or extracts of them, upon a special search made by order of the Roman Senate, including those already mentioned as of the time of Tarquin (some nine hundred years earlier), were gathered up in Greece, Italy and other parts and destroyed, because, as we are informed, this Roman Senate had embraced the Christian faith and furthermore "such vanities began to grow out of fashion; till at last Stilicho burnt them all under Honorius (a son of Theodosius the Great), for which he is so severely censured by the noble poet Rutilius, in his ingenious itinerary."

Not only Roman Arms the Wretch betrayed
To barbarous Foes; before that cursed Deed,
He burnt the Writings of the sacred Maid,
We hate Althaea for the fatal Brand;
When Nisius fell, the weeping Birds complained:

More cruel he than the revengeful Fair;
More cruel heth at Nisius' Murderer.
Whose impious Hands into the Flames have thrown
The Heavenly Pledges of the Roman Crown,
Unrav'lling all the Doom that careful Fate had spun."

The destruction of Rome by Alaric, King of the Western Goths, A. D. 410, and the subsequent dismemberment of the entire Roman Empire by the barbarians of the North who followed in his wake, announced that ancient history had come to an end.

It may be truly said as well that the ending of the ancient history of the black and colored writing inks which began in the obscurity of tradition between 2000 and 1800 B. C., a period of some 2200 years, was also contemporaneous with these events.

The eclipse of ink-written literature for at least 500 of the 1000 years which followed, and known as the Middle or "Dark" Ages, except in the Church alone, who seem to have kept up the production of manuscript books principally for ecclesiastical and medical purposes was complete. Hence, any information pertaining to those epochs about ink, writing materials and ink writings, must be sought for in the undestroyed records and the ink writings themselves left by the fathers of the Church. All else is tainted and of doubtful authority.

* * * * * * * *

"When waned the star of Greece was there no cry,
To rouse her people from their lethargy?
Was there no sentry on the Parthenon—
No watch-fire on the field of Marathon,
When science left the Athenian city's gate,
To seek protection from a nameless fate?
The sluggish sentry slept—no cry was heard
No hands the glimm'ring watch-fire's embers stirr'd.
Fair science unmolested left the land,
That she had nurtured with maternal hand;
And wandered forth some genial spot to find,
Where she might rear her altar to the mind.
"Long thro' the darken'd ages of a world,

Back to primeval chaos rudely hurled,
She journey'd on amid the gath'ring gloom,
A spectre form emerging from the tomb.
Earth had no resting place—no worshipper—
No dove returned with olive branch to her:
Her lamp burned dimly, yet its flick'ring light,
Guided the wanderer thro' the lengthen'd night.
Oft in her weary search, she paused the while,
To catch one gleam of hope—one favour'd smile;
But the dim mists of ignorance still threw,
Their blighting influence o'er the famish'd few,
Who deigned to look upon that lustrous eye,
Which pierced the ages of futurity.

"For ten long centuries she groped her way,
Through gloom, and darkness, ruin and decay;
Yet came at last the morning's rosy light,
A thousand echoes hail'd the glorious sight—
Joy thrill'd the universe—one iningled cry
Of exultation, pealed along the sky!
Science came forth in richer robes arrayed
She trod a pathway ne'er before essayed;
Up the steep mount of fame she fleetly pressed,
And hung her trophies on its gilded crest."

CHAPTER IV.

CLASSICAL INK AND ITS EXODUS (CONTINUED).

DESTRUCTION OF THE PERGAMUS LIBRARY OF ALEXANDRIA— SOME OBSERVATIONS BY SIR THOMAS ASTLE — COMPARISON OF HIS STATEMENTS WITH THOSE OF PROFESSOR ANTHON RELATIVE TO FRAGMENTS OF ANTIQUITY WHICH REMAIN—AUTHENTICITY OF THEM NOT DISTURBED IF THEY ARE OF PROPER AGE —TAYLOR'S VIEWS ON THIS SUBJECT.

THE storming of Alexandria and the destruction of the Pergamus library, composed largely of ink-written volumes, by the Saracens, A. D. 642, has already been reverted to. Astle observes:

"Thus perished by fanatical madness, the inestimable Alexandrian library, which is said to have contained at that time upwards of five hundred thousand volumes; and from this period, barbarity and ignorance prevailed for several centuries. In Italy and all over the west of Europe learning was in a measure extinguished, except some small remains which were preserved in Constantinople.

"Theodosious, the younger, was very assiduous in augmenting this library, by whom, in the latter end of the fourth century, it was enlarged to one hundred thousand volumes, above one-half of which were burnt in the fifth century by the Emperor Leo the First, so famous for his hatred to images.

"The inhabitants of Constantinople had not lost their taste for literature in the beginning of the thirteenth century, when this city was sacked by the Crusaders, in the year 1205; the depredations then committed are related in Mr. Harris's posthumous works, vol. ii, p. 301, from Nicetas the Choniate, who was present at the sacking of this place. His account of the statues, bustos, bronzes, manuscripts, and other exquisite remains of antiquity, which then perished, cannot be read by any lover of arts and learning without emotion.

"The ravages committed by the Turks who plundered Constantinople, in the year 1453, are related by Philelphus, who was a man of learning, and was tutor to aeneas Sylvius (afterwards pope, under the name of Pius the Second) and was an eye-witness to what passed at that time. This tutor says, that the persons of quality, especially the women, still preserved the Greek language uncorrupted. He observes, that though the city had been taken before, it never suffered so much as at that time; and adds, that, till that period, the remembrance of ancient wisdom remained at Constantinople, and that no one among the Latins was deemed sufficiently learned, who had riot studied for some time at that place; he expressed his fear that all the works of the ancients would be destroyed.

"Still, however, there are the remains of three libraries at Constantinople: the first is called that of Constantine the Great; the second is for all ranks of people without distinction; the third is in the palace, and is called the Ottoman library; but a fire consumed a great part of the palace, and almost the whole library, when as is supposed, Livy and a great many valuable works of the ancients perished. Father Possevius has given an account of the libraries at Constantinople, and in other parts of the Turkish dominions, in his excellent work entitled, Apparatus Sacer. (He calls attention to no less than six thousand authors.)

Many other losses of the writings of the ancients have been attributed to the zeal of the Christians, who at different periods made great havock amongst the Heathen authors. Not a single copy of the work of Celsus is now to be found, and what we know of that work is from Origen, his opponent. The venerable fathers, who employed themselves in erasing the best works of the most eminent Greek or Latin authors, in order to transcribe the lives of saints or legendary tales upon the obliterated vellum, possible mistook these lamentable depredations for works of piety. The ancient fragment of the 91st book of Livy, discovered by Mr. Bruns, in the Vatican, in 1772, was much defaced by the pious labours of some well-intentioned divine. The Monks made war on books as the Goths had done before them. Great numbers of manuscripts have also been destroyed in this kingdom (Great Britain) by its invaders, the Pagan Danes, and the Normans, by the civil commotions raised by the barons, by the bloody contests between the houses of York and Lancaster, and especially by the general plunder and devastations of monasteries and religious houses in the reign of Henry the Eighth; by the ravages committed in the civil war in the time of Charles the First, and by the fire that happened in the Cottonian library, October 23, 1731."

Mr. Astle's comments on the volumes or remnants of volumes which remain to us, becomes most interesting in the lights thrown on them by Professor Anthon in his "Classical Dictionary," 1841, which are quoted in part following those of Mr. Astle.

Mr. Astle remarks:

"The history of Phoenicia by Sanconiatho, who was a contemporary with Solomon, would have been entirely lost to us, had it not been for the valuable fragments preserved by Eusebius."

Says Prof. Anthon:

"Sanchoniathon, a Phoenician author, who if the fragments of his works that have reached us be genuine, and if such a person ever existed, must be regarded as the most ancient writer of whom we have any knowledge after Moses. As to the period when be flourished, all is uncertain. He is the author of three principal works, which were written in Phoenician. They were translated into the Greek language by Herennius Philo, who lived in the second century of our era. It is from this translation which we obtain all the fragments of Sanchoniathon that have reached our times. Philo had divided his translation into nine books, of which Porphyry made use in his diatribe against the Christians. It is from the fourth book of this lost work that Eusebius took, for an end directly opposite to this, the passages which have come down to us. And thus we have those documents relating to the mythology and history of the Phoenicians from the fourth hand."

Mr. Astle continues:

"Manetho's History of Egypt, and the History of Chaldea, by Berosus, have nearly met with the same fate."

From Anthon:

"Berosus; a Babylonian historian. He was a priest of the temple of Belus in the time of Alexander. The ancients mention three books of his of which Josephus and Eusebius have preserved fragments. Annius of Viterbo published a work under the name of Berosus, which was soon discovered to be a forgery."

By Astle:

"The Historical Library of Diodorus Siculus consisted likewise of forty books, but only fifteen are now extant; that is, five between the fifth and the eleventh, and the last ten, with some fragments collected out of Photius and others."

By Anthon:

"Diodorus, surnamed Siculus, a contemporary of Julius Caesar and Agustus. He published a general history in forty books, under the title 'Historical Library,' which covered a period of 1138 years. We have only a small part remaining of this vast compilation. These rescued portions we owe to Eusebius, to John Malala and other writers of the lower empire, who have cited them in the course of their works. He is the reputed author of the famous sophism against motion. 'If any body be moved, it is moved in the place where it is, or in a place where it is not, for nothing can act or suffer where it is not, and therefore there is no such thing as motion.' "

By Astle:

"The General History of Polybius originally contained forty books; but the first five only, with some extracts or fragments, are transmitted to us."

By Anthon:

"Polybius, an eminent Greek historian, born about, B. C. 203. Polybius gave to the world various historical writings, which are entirely lost with the exception of his General History. It embraced a period of 53 years. Of the forty books which it originally comprehended, time has spared only the first five entire. Of the rest, as far as the seventeenth, we have merely fragments though of considerable size. Of the remaining books we have nothing left except what is found in two merger abridgments which the Emperor Constantine Porphyrogenitus, in the tenth century caused to be made of the whole work."

From Astle:

"Dionysius Halicarnassensis wrote twenty books of Roman antiquities, extending from the siege of Troy, to the Punic war A. U. C.

488; but only eleven of them are now remaining, which reach no further than the year of Rome 312."

From Anthon:

"He was born in the first century B. C. His principal work was 'Roman Antiquities.' It originally consisted of twenty books, of which the first ten remain entire. Dionysius wrote for the Greeks, and his object was to relieve them from the mortification which they felt at being conquered by a race of barbarians, as they considered the Romans to be. And this he endeavored to effect by twisting and forging testimonies, and botching up the old legends, so as to make out a prima facie proof of the Greek origin of the city of Rome. Valuable additions were made in 1816, by Mai, from an old MSS."

By Astle:

"Appian is said to have written the Roman History in twenty-four books; but the greatest part of the works of that author is lost."

By Anthon:

"He was the author of a Roman History in twenty-four books which no longer exist entire; the parts missing have been supplied but was not written by Appian but is a mere compilation from Plutarch's Lives of Crassus and Antony."

By Astle:

"Dion Cassius wrote eighty books of history, but only twenty-five are remaining, with some fragments, and an epitome of the last twenty by Xiphilinus."

By Anthon:

"His true name was Cassius, born A. D. 155; —we have fragments remaining of the first thirty- six books, they comprehend a period from B. C. 65 to B. C. 10;—they were found by Mai in two Vatican MSS., which contain a sylloge or collection made by Maximus Planudes (who lived in the fourteenth century. He was the first Greek that made use of the Arabic numerals as they are called)."

Mr. Astle further observes:

"The Emperor Tacitus ordered ten copies of the works of his relation, the historian, to be made every year which he sent into the different provinces of the empire; and yet, notwithstanding his endeavours to perpetuate these inestimable works, they were buried in oblivion for many centuries. Since the restoration of learning an ancient MSS. was discovered in a monastery in Westphalia, which contained the most valuable part of his annals; but in this unique manuscript, part of the fifth, seventh, ninth and tenth books are deficient, as are part of the eleventh, and the latter part of the sixteenth. This MSS. was procured by that great restorer of learning Pope Leo X., under whose patronage it was printed at Rome in 1515; he afterwards deposited it in the Vatican library, where it is still preserved. Thus posterity is probably indebted to the above magnificent Pontiff, for the most valuable part of the works of this inimitable historian."

Accounts which differentiate in their descriptive details of questioned ink-written fragments of antiquity and on the genuineness or authenticity of which rests the truth or falsity of ancient history or other literature, serve to taint such remains with a certain degree of suspicion and doubt. When, however, in the light of investigation, the materials of which they are composed are found to approach closely the age they purport to represent, then it is that such fragments can be said to have fairly established their own identity.

Taylor asserts:

"The remote antiquity of a manuscript is of ten established by the peculiar circumstance of its existing BENEATH another writing. Some invaluable manuscripts of the Holy Scriptures, and not a few precious fragments of classic literature, have been thus brought to light.

"The age of a manuscript may often be ascertained with little chance of error, by some such indications as the following:—the quality or appearance of the INK, the nature of the material; that is to say, whether it be soft leather, or parchment, or the papyrus of Egypt, or the bombycine paper; for these materials succeeded each other, in common use, at periods that are well known;— the peculiar form, size, and character of the writing; for a regular progression in the modes of writing may be traced by abundant evidence through every age from the remotest times;—the style of the ornaments or illuminations, as they are termed, often serves to indicate the age of the book which they decorate.

"From such indications as these, more or less definite and certain, ancient manuscripts, now extant, are assigned to various periods, extending from the sixteenth, to the fourth century of the Christian era; or perhaps, in one or two instances, to the third or second. Very few can claim an antiquity so high as the fourth century; but not a few are safely attributed to the seventh; and a great proportion of those extant were unquestionably executed in the tenth; while many belong to the following four hundred years. It is, however, to be observed, that some manuscripts, executed at so late a time as the thirteenth, or even the fifteenth century, afford clear internal evidence that, by a single remove only, the text they contain claims a REAL antiquity, higher than that even of the oldest existing copy of the same work. For these older copies sometimes prove, by the peculiar nature of the corruptions which have crept into the text, that they have been derived through a long series of copies; while perhaps the text of the more modern manuscripts possesses such a degree of purity and freedom from all the usual consequences of frequent transcription, as to make it manifest that the copy from which it was taken, was so ancient as not to be far distant from the time of the first publication of the work."

CHAPTER V.

REVIVAL OF INK.

THE DISAPPEARANCE AND PRESERVATION OF INK WRITINGS, AS ESTIMATED BY LA CROIX—COMMENTS OF OTHER WRITERS—DE VINNE'S INTERESTING EXPLANATIONS OF THE STATUS QUO OF MANUSCRIPT WRITINGS DURING THE DARK AGES WHICH PRECEDED THE INVENTION OF PRINTING—PRICES PAID FOR BOOKS IN ANCIENT TIMES—LIMITATIONS OF HANDWRITING AND HANDWRITING MATERIALS AT THE BEGINNING OF THE FIFTH CENTURY—WHO CONTROLLED THE RECORDS ABOUT THEM—

INVENTION OF THE QUILL PEN—THE CAUSE OF INCREASED FLUIDITY OF INKS—ORIGIN OF THE SECRETA—CHARACTER OF INFORMATION OBTAINED FROM THEM—IMPROVEMENT OF BLACK INKS IN THE EIGHTH CENTURY AND EMPLOYMENT OF POMEGRANITE INK.

LA CROIX' preface to his "Science and Literature of the Middle Ages and the Renaissance," refers to the Dark Ages:

"In the beginning of the Middle Ages, at the commencement of the fifth century, the Barbarians made an inroad upon the old world; their renewed invasions crushed out, in the course of a few years, the Greek and Roman civilization; and everywhere darkness succeeded to light. The religion of Jesus Christ was alone capable of resisting this barbarian invasion, and science and literature, together with the arts, disappeared from the face of the earth, taking refuge in the churches and monasteries. It was there that they were preserved as a sacred deposit, and it was thence that they emerged when Christianity had renovated pagan society. But centuries and centuries elapsed before the sum of human knowledge was equal to what it had been at the fall of the Roman empire. A new society, moreover, was needed for the new efforts of human intelligence as it resumed its rights. Schools and universities were founded under the auspices of the clergy and of the religious corporations, and thus science and literature were enabled to emerge from their tombs. Europe, amidst the tumultuous conflicts of the policy which made and unmade kingdoms, witnessed a general revival of the scholastic zeal; poets, orators, novelists, and writers increased in numbers and grew in favour; savants, philosophers, chemists and alchemists, mathematicians and astronomers, travellers and naturalists, were awakened, so to speak, by the life-giving breath of the Middle Ages; and great scientific discoveries and admirable works on every imaginable subject showed that the genius of modern society was not a whit inferior to that of antiquity. Printing, was invented, and with that brilliant discovery, the Middle Ages, which had accomplished their work of social renovation, made way for the Renaissance, which scattered abroad in profusion the prolific and brilliant creations of Art, Science, and Literature."

This author to some extent discredits himself, however, p. 455, where he remarks:

"Long before the invasions of the Barbarians the histories written by Greek and Latin authors concerning the annals of the ancient peoples had been falling into disfavor. Even the best of them were little read, for the Christians felt but slight interest in these pagan narratives, and that is why works relating to the history of antiquity were already so scarce."

Another authority writing on the same subject discusses it from a different standpoint, remarking:

"As in the middle ages invention busied itself with instruments of torture, and as in our days it is taken up almost as much with the destructive engines of war as with the productive arts of peace, so in those early ages it applied itself to the fabrication of idols, to the mechanism and theatrical contrivances for mysteries and religious ceremonies. There was then no desire to communicate discoveries, science was a sort of freemasonry, and silence was effectually secured by priestly anathemas; men of science were as jealous of one another as they were of all other classes of society. If we wish to form a clear picture of this earliest stage of civilization, an age which represents at once the naivete of childhood and the suspicious reticence of senility, we must turn our eyes to the priest, on the one hand, claiming as his own all art and science, and commanding respect by his contemptuous silence; and, on the other hand, to the mechanic plying the loom, extracting the Tyrian dye, practising chemistry, though ignorant of its very name, despised and oppressed, and only tolerated when he furnished Religion with her trappings or War with arms. Thus the growth of chemistry was slow, and by reason of its backwardness it was longer than any other art in ridding itself of the leading-strings of magic and astrology. Practical discoveries must have been made many times without science acquiring thereby any new fact. For to prevent a new discovery from being lost there must be such a combination of favorable circumstances as was rare in that age and for many succeeding ages. There must be publicity, and publicity is of quite recent growth; the application of the discovery must be not only possible but obvious, as satisfying some want. But wants are only felt as civilization progresses. Nor is that all; for a practical discovery to become a scientific fact it must serve to demonstrate the error of one hypothesis, and to suggest a new one, better fitted for the synthesis of existing facts. But (some) old beliefs are proverbially obstinate and virulent in their opposition to newer and truer theories which are destined to eject and replace them. To sum up, even in our own day, chemistry rests on a less sound basis than either physics, which had the advantage of originating as late as the 17th century, or astronomy, which dates from the time when the Chaldean shepherd had sufficiently provided for his daily wants to find leisure for gazing into the starry Heavens."

The observations of a still earlier commentator are of the same general nature. He says:

"In the first ages of Christianity, when the fathers of the Church, the Jews, and the Heathen philosophers were so warmly engaged in controversy, there is reason to believe that pious frauds were not uncommon: and that when one party suspected forgeries, instead of an attempt at confutation, which might have been difficult, they had recourse perhaps to a countermine: and either invented altogether, or eked out some obscure traditional scraps by the embellishments of fancy. When we consider, amongst many literary impositions of later times, that Psalmanazar's history of Formosa was, even in this enlightened age and country (England, about 1735), considered by our most learned men as unquestionably authentic, till the confession of the author discovered the secret, I think it is not difficult to conceive how forgeries of remote events, before the invention of printing and the general diffusion of knowledge might gain an authority, and especially with the zealous, hardly inferior to that of the most genuine history."

De Vinne, however, in his "Invention of Printing," New York, 1878, best explains the status quo of those times, relative not only to book (MSS.) making, and methods of circulation, but the causes which led up to their eventual disappearance and the literary darkness which ensued. His remarks are so pertinent that they are quoted at length:

"The civilization of ancient Rome did not require printing. If all the processes of typography had been revealed to its scholars the art would not have been used. The wants of readers and writers were abundantly supplied by the pen. Papyrus paper was cheap, and scribes were numerous; Rome had more booksellers than it needed, and books were made faster than they could be sold. The professional scribes were educated slaves, who, fed and clothed at nominal expense, and organized under the direction of wealthy publishers, were made so efficient in the production of books, that typography, in an open competition, could have offered few advantages.

"Our knowledge of the Roman organization of labor in the field of bookmaking is not as precise as could be wished; but the frequent notices of books, copyists and publishers, made by many authors during the first century, teach us that books were plentiful. Horace, the elegant and fastidious man of letters, complained that his books were too common, and that they were sometimes found in the hands of vulgar snobs for whose entertainment they were not written. Martial, the jovial man of the world, boasted that his books of stinging epigrams were to be found in everybody's hands or pockets. Books were read not only in the libraries, but at the baths, in the porticoes of houses, at private dinners and in mixed assemblies. The business of bookmaking was practised by too many people, and some were incompetent. Lucian, who had a keen perception of pretense in every form, ridicules the publishers as ignoramuses. Strabo, who probably wrote illegibly, says that the books of booksellers were incorrect.

"The price of books made by slave labor was necessarily low. Martial says that his first book of epigrams was sold in plain binding for six sesterces, about twenty-four cents of American money; the same book in sumptuous binding was valued at five denarii, about eighty cents. He subsequently complained that his thirteenth book was sold for only four sesterces, about sixteen cents. He frankly admits that half of this sum was profit, but intimates, somewhat ungraciously, that the publisher Tryphon gave him too small a share. Of the merits of this old disagreement between the author and publisher we have not enough of facts to justify an opinion. We learn that some publishers, like Tryphon and the brothers Sosii, acquired wealth, but there are many indications that publishing was then, as it is now, one of the most speculative kinds of business. One writer chuckles over the unkind fate that sent so many of the unsold books of rival authors from the warehouses of the publisher, to the shops of grocers and bakers, where they were used to wrap up pastry and spices; another writer says that the unsold stock of a bookseller was sometimes bought by butchers and trunk makers.

"The Romans not only had plenty of books but they had a manuscript daily newspaper, the Acta Diurna, which seems to have been a record of the proceedings of the senate. We do not know how it was written, nor how it was published, but it was frequently mentioned by contemporary writers as the regular official medium for transmitting intelligence. It was sent to subscribers in distant cities, and was, sometimes, read to an assembled army. Cicero mentions the Acta as a sheet in which he expected to find the city news and gossip about marriages and divorces.

"With the decline of power in the Roman empire came the decline of literature throughout the world. In the sixth century the business of bookmaking had fallen into hopeless decay. The books that had been written were seldom read, and the number of readers diminished with every succeeding generation. Ignorance pervaded in all ranks of society. The Emperor Justin I, who reigned between the years 518 and 527, could not write, and was obliged to sign state papers with the form of stencil plate that had been recommended by Quintilian. Respect for literature was dead. In the year, 476, Zeno, the Isaurian, burned 120,000 volumes in the city of Constantinople. During the year 640, Amrou, the Saracen, fed the baths of Alexandria for six months with the 500,000 books that had been accumulating for centuries in its famous library of the Serapion. Yet books were so scarce in Rome at the close of the seventh century that Pope Martin requested one of his bishops to supply them, if possible, from Germany. The ignorance of ecclesiastics in high station was alarming. During this century, and for centuries afterward, there were many bishops and archbishops of the church who could not sign their names. It was asserted at a council of the church held in the year 992, that scarcely a single person was to be found in Rome itself who knew the first elements of letters. Hallam says, 'To sum up the account of ignorance in a word, it was rare for a layman of any rank to know bow to sign his name.' He repeats the statements that Charlemagne could not write, and Frederic Barbarossa could not read. John, king of Bohemia, and Philip, the Hardy, king of France, were ignorant of both accomplishments. The graces of literature were tolerated only in the ranks of the clergy; the layman who preferred letters to arms was regarded as a man of mean spirit. When the Crusaders took Constantinople, in 1204, they exposed to public ridicule the pens and inkstands that they found in the conquered city as the ignoble arms of a contemptible race of students.

"During this period of intellectual darkness, which lasted from the fifth until the fifteenth century, a period sometimes described, and not improperly, as the dark ages, there was no need for any improvement in the old method of making books. The world was not then ready for typography. The invention waited for readers more than it did for types; the multitude of book buyers upon which its success depended had to be created. Books were needed as well as readers. The treatises of the old Roman sophists and rhetoricians, the dialectics of Aristotle and the schoolmen, and the commentaries on ecclesiastical law of the fathers of the church, were the works which engrossed the attention of men of letters for many centuries before the invention of typography. Useful as these books may have been to the small class of readers for whose benefit they were written, they were of no use to a people who needed the elements of knowledge."

In the more ancient times, however, when MSS. books (rolls) were not quite so plentiful there was seemingly no difficulty in obtaining large sums for them.

Aristotle, died B. C. 322, paid for a few books of Leusippus, the philosopher, three Attick talents, which is about $3,000. Ptolemy Philadelphus is said to have given the Athenians fifteen talents, an exemption from tribute and a large supply of provisions for the MSS. of aeschylus, Sophocles and Euripides written by themselves.

Arbuthnot, discussing this subject, remarks that Cicero's head, "which should justly come into the account of Eloquence brought twenty-five Myriads of Drachms, which is the equivalent of $40,000. Also, "the prices of the magical books mentioned to be burnt in the Acts of the Apostles is five. Myriads of Pieces of Silver or Drachms."

Picolimini relates that the equivalent of eighty golden crowns was demanded for a small part of the works of Plutarch.

If we are to believe any of the accounts, the environment of the art of handwriting and handwriting materials at the beginning of the fifth century had contracted within a small compass, due principally to the general ignorance of the times.

As practiced it was pretty much under the control of the different religious denominations and the information obtainable about inks from these sources is but fragmentary. What has come down to us of this particular era is mostly found on the old written Hebrew relics, showing that they at least had made no innovations in respect to the use of their ritualistic deyo.

The invention of the quill pen in the sixth century permitted a degree of latitude in writing never before known, the inks were made thinner and necessarily were less durable in character. Greater attention was given to the study and practice of medicine and alchemy which were limited to the walls of the cloister and secret places. The monk physicians endeavored by oral instructions and later by written ones to communicate their ink-making methods not only of the black and colored, but of secret or sympathetic inks, to their younger brethren, that they might thus be perpetuated. All the traditional and practical knowledge they possessed was condensed into manuscript forms; additions from other hands which included numerous chemical receipts for dyeing caused them to multiply; so that as occasion required from time to time, they were bound up together booklike and then circulated among favored secular individuals, under the name of "Secreta."

The more remote of such treatises which have come down to us seem to indicate the trend of the researches respecting what must have been in those times unsatisfactory inks. Scattered through them appear a variety of formulas which specify pyrites (a combination of sulphur and metal), metals, stones and other minerals, soot, (blue) vitriol, calxes (lime or chalk), dye-woods, berries, plants, and animal colors, some of which if made into ink could only have been used with disastrous results, when permanency is considered.

The black ink formulas of the eighth century are but few, and show marked improvement in respect to the constituents they call for, indicating that many of those of earlier times had been tried and found wanting. One in particular is worthy of notice as it names (blue) vitriol, yeast, the lees (dregs) of wine and the rind of the pomegranate apple, which if commingled together would give results not altogether unlike the characteristic phenomena of "gall" ink. Confirmation of the employment of such an ink on a document of the reign of Charlemigne in the beginning of the ninth century on yellow-brown Esparto (a Spanish rush) paper, is still preserved. Specimens of "pomegranate" ink, to which lampblack and other pigments had been added of varying degrees of blackness, on MSS., but lessening in number as late as the fourteenth century, are still extant in the British Museum and other public libraries.

CHAPTER VI.

INK OF THE WEST.

REMARKS OF ARCH-DEACON CARLISLE—WHEN READING AND WRITING CEASED TO BE MYSTERIES—ORIGIN OF THE WORDS CLERK AND SIGN—SCARCITY OF MANU-SCRIPTS —FOUNDING OF IRISH SCHOOLS OF LEARNING IN THE SEVENTH CENTURY—MONKS NOT PERMITTED TO USE ARTIFICIAL LIGHT IN PREPARING MSS.—OBSERVATIONS OF MADAN ABOUT THE HISTORY OF WRITING DURING THE DARK AGES—INK- WRITTEN MSS. TREASURES.

THE ancient history of the art of writing in more northern sections of the Western world, William Nicolson, Arch-Deacon of Carlisle, author of "The English Historical Library," London, 1696, tells very quaintly:

"The Danes register'd their more considerable transactions upon Rocks; or on parts of them, hewen into various Shapes and Figures. On these they engrav'd such Inscriptions as were proper for their Heathen Alters, Triumphal Arches, Sepulchral Monuments and Genealogical Histories of their Ancestors. Their writings of less concern (as Letters, Almanacks, &c.) were engraven upon Wood: And because Beech was most plentiful in Demnark, (tho Firr and Oak be so in Norway and Sweden) and most commonly employ'd in these Services, form the word Bog (which in their Language is the Name of that sort of Wood) they and all other Northern Nations have the Name of Book. The poorer sort used Bark; and the Horns of Rain- Deer and Elks were often finely polish'd and shaped into Books of several Leaves. Many of these old Calendars are likewise upon Bones of Beasts and Fishes: But the Inscriptions on Tapestry, Bells, Parchment and Paper, are of later use.

"Some other Monuments may be known to be of a Danish Extraction, tho they carry nothing of a Runic Inscription. Few of their Temples were cover'd; and the largest observ'd by Wormius (at Kialernes in Island) was 120 foot in length, and 60 in breadth.

"The next Monument of Age is their Edda Islandorum; the meaning of which Appellation they that publish the Book hardly pretend to understand. As far as I can give the Reader any satisfaction, he is to. know that Island was first inhabited (in the year 874) by a Colony of Norwegians; who brought hither the Traditions of their Forefathers, in certain metrical Composures, which (as is usual with Men transplanted into a Foreign Land) were here more zealously and carefully preserv'd and kept in memory than by the Men of Norway themselves. About 240 years after this (A. D. 1114) their History began to be written by one Saemund, surnam'd Frode or the wise; who (in nine years' travel through Italy, Germany and England) had amass'd together a mighty Collection of Historical Treatises. With these he return'd full fraught into Island; where he also drew up an account of the affairs of his own Country. Many of his Works are now said to be lost: But there is still an Edda, consisting of several Odes (whence I suspect its Name is derived) written by many several hands, and at different times, which bears his Name. The Book is a Collection of Mythological Fables, relating to the ancient State and Behaviour of the Great Woden and his followers, in terms poetical and adapted to the Service of those that were employ'd in the composure of their old Rhymes and Sonnets.

"There is likewise extant a couple of Norwegian Histories of good Authentic Credit; which explains a great many particulars relating to the Exploits of the Danish Kings in Great Britain, which our own Historians have either wholly omitted or very darkly recorded. The former of these was written soon after the year 1130, by one Theodoric a Monk, who acknowledges his whole Fabrick to be built upon

Tradition, and that the old Northern History is no where now to be had save only ab Islendingorum antiquis Carminibus.

" 'Tis a very discouraging Censure which Sir William Temple passes upon all the Accounts given us of the Affairs of this Island, before the Romans came and Invaded it. The Tales (says he) we have of what pass'd before Caesar's Time, of Brute and his Trojans, of many Adventures and Successions, are cover'd with the Rust of Time, or Involv'd in the Vanity of Fables or pretended Traditions; which seem to all Men obscure or uncertain, but to be forged at pleasure by the Wit or Folly of their first Authors, and not to be regarded. And again; I know few ancient Authors upon this Subject (of the British History) worth the pains of perusal, and of Dividing or Refining so little Gold out of so much course Oar, or from so much Dross. But some other Inferiour People may think this worth their pains; since all Men are not born to be Ambassadors: And, accordingly, we are told of a very Eminent Antiquary who has thought fit to give his Labours in this kind the Title of Aurum, ex Stercore. There's a deal of Servile Drudgery requir'd to the Discovery of these riches, and such as every Body will not stoop to: for few Statesmen and Courtiers (as one is lately said to have observ'd in his own Case) care for travelling in Ireland, or Wales, purely to learn the Language.

"A diligent Enquirer into our old British Antiquities would rather observe (with Industrious Leland) that the poor Britains, being harass'd by those Roman Conquerours with continual Wars, could neither have leisure nor thought for the penning of a Regular History: and that afterwards their Back-Friends, the Saxons, were (for a good while) an Illiterate Generation; and minded nothing but Killing and taking Possession. So that 'tis a wonder that even so much remains of the Story of those Times as the sorry Fragments of Gildas; who appears to have written in such a Consternation, that what he has left us looks more like the Declamation of an Orator, hired to expose the miserable Wretches, than any Historical Account of their Sufferings."

Palgrave asserts that reading and writing were no longer mysteries after the pagan age, but were still acquirements almost wholly confined to the clergy.

The word "clericus" or "clerk," became synonymous with penman, the sense in which it is still most usually employed. If a man could write, or even read, his knowledge was considered as proof presumptive that he was in holy orders. If kings and great men had occasion to authenticate any document, they subscribed the "sign" of the cross opposite to the place where the "clerk" had written their name. Hence we say, to sign a deed or a letter.

Books (MSS.) were extremely rare amongst the Scandinavian and northern nations. Before their communication with the Latin missionaries, wood appears to have been the material upon which their runes were chiefly written: and the verb "write," which is derived from a Teutonic root, signifying to scratch or tear, is one of the testimonies of the usage. Their poems were graven upon small staves or rods, one line upon each face of the rod; and the Old English word "stave," as applied to a stanza, is probably a relic of the practice, which, in the early ages, prevailed in the West. Vellum or parchment afterwards supplied the place of these materials. Real paper, manufactured from the pellicle of the Egyptian reed or papyras, was still used occasionally in Italy, but it was seldom exported to the countries beyond the Alps; and the elaborate preparation of the vellum, upon which much greater care was bestowed than in the modern manufacture, rendered it a costly article; so much so, that a painstaking clerk could find it worth his while to erase the writing of an old book, in order to use the blank pages for another manuscript. The books thus rewritten were called "codices rescripti," or "palimpsests." The evanescent traces of the first layer of characters may occasionally be discerned beneath the more recent text which has been imposed upon them.

In Ireland, first known as the Isle of Saints, was founded in the seventh century a great school of learning which included writing and illuminating, which passed to the English by way of the monasteries created by Irish monks in Scotland. Their earliest existing MSS. are said to belong to that period. In the Irish scriptoriums (rooms or cells for writing) of the Benedictine monasteries where they were prepared, so particular were the monks that the scribes were forbidden to use artificial light for fear of injuring the manuscripts.

Most interesting and entertaining are the observations of Falconer Madan, a modern scholar of some repute. Of the history of writing in ink during the "Dark Ages" he says:

"In the seventh and eighth centuries we find the first tendency to form national hands, resulting in the Merovingian or Frankish hand, the Lombardic of Italy, and the Visigothic of Spain. These are the first difficult bands which we encounter; and when we remember that the object of writing is to be clear and distinct, and that the test of a good style is that it seizes on the essential points in which letters differ, and puts aside the flourishes and ornaments which disguise the simple form, we shall see how much a strong influence was needed to prevent writing from becoming obscure and degraded. That influence was found in Charles the Great.

"In the field of writing it has been granted to no person but Charles the Great to influence profoundly the history of the alphabet. With rare insight and rarer taste he discountenanced the prevalent Merovingian hand, and substituted in eclectic hand, known as the Carolingian Minuscule, which way still be regarded as a model of clearness and elegance. The chief instrument in this reform was Alcuin of York, whom Charles placed, partly for this purpose, at the head of the School of Tours in A. D. 796. The selection of an Englishman for the post naturally leads us to inquire what hands were then used in England, and what amount of English influence the Carolingian Minuscule, the foundation of our modern styles, exhibits.

"If we gaze in wonder on the personal influence of Charles the Great in reforming handwriting, we shall be still more struck by the spectacle presented to us by Ireland in the sixth, seventh and eighth centuries. It is the great marvel in the history of writing. Modern historians have at last appreciated the blaze of life, religions, literary, and artistic, which was kindled in the 'Isle of Saints' within a century after St. Patrick's coming (about A. D. 450); how the enthusiasm kindled by Christianity in the Celtic nature so far transcended the limits of the island, and indeed of Great Britain, that Irish missionaries and monks were soon found in the chief religious centres of Gaul, Germany, Switzerland, and North Italy, while foreigners found their toilsome way to Ireland to learn Greek! But less prominence has been given to the artistic side of this great reflex movement from West to East than to the other two. The simple facts attest that in the seventh century, when our earliest existing Irish MSS. were written, we find not only a style of writing (or indeed two) distinctive, national, and of a high type of excellence, but also a school of illumination which, in the combined lines of mechanical accuracy and intricacy, of fertile invention of form and figure and of striking arrangements of colour, has never been surpassed. And this is in the seventh century—the nadir of the rest of Europe!

"It is certain that Alcuin was trained in Hiberno- Saxon calligraphy, so that we may be surprised to find that the writing which, under Charles the Great, he developed at Tours, bears hardly a trace of the style to which he was accustomed. En revanche, in the ornamentation and illumination of the great Carolingian volumes which have come down to our times, we find those constant, persistent traces of English and Irish work which we seek for in vain in the plainer writing.

"This minuscule superseded all others almost throughout the empire of Charles the Great, and during the ninth, tenth, and eleventh centuries underwent very little modification. Even in the two next centuries, though it is subject to general modification, national differences are hardly observable, and we can only distinguish two large divisions, the group of Northern Europe (England, North France, Italy, and Spain). The two exceptions are, that Germany, both in writing and painting, has always stood apart, and lags behind the other nations of Western Europe in its development, and that England retains her Hiberno-Saxon hand till after the Conquest of 1066. It may be noted that the twelfth century produced the finest writing ever known—a large, free and flowing form of the minuscule of Tours. In the next century comes in the angular Gothic hand, the difference between which and the twelfth century hand may be fairly understood by a comparison of ordinary German and Roman type. In the thirteenth, fourteenth, and fifteenth centuries the writing of each century may be discerned, while the general tendency is towards complication, use of abbreviations and contractions, and development of unessential parasitic forms of letters.

"The Book of Kells, the chief treasure of Trinity College, Dublin, is so-called from having been long preserved at the Monastery of Kells, founded by Columba himself. Stolen from thence, it eventually passed into Archbishop Ussher's hands, and, with other parts of his library, to Dublin. The volume contains the Four Gospels in Latin, ornamented with extraordinary freedom, elaboration, and beauty. Written apparently in the seventh century, it exhibits, both in form and colour, all the signs of the full development and maturity of the Irish style, and must of necessity have been preceded by several generations of artistic workers, who founded and improved this particular school of art. The following words of Professor Westwood, who first drew attention to the peculiar excellences of this volume, will justify tile terms made use of above: 'This copy of the Gospels, traditionally asserted to have belonged to Columba, is unquestionably the most elaborately executed MS. of early art now in existence, far excelling, in the gigantic size of the letters in the frontispieces of the Gospel, the excessive minuteness of the ornamental details, the number of its decorations, the fineness of the writing, and the endless variety of initial capital letters with which every page is ornamented, the famous Gospels of Lindisfarne in the Cottonian Library. But this MS. is still more valuable on account of the various pictorial representations of different scenes in the life of our Saviour, delineated in a style totally unlike that of every other school.' "

CHAPTER VII.

EARLY MEDIAEVAL INK.

CONTROVERSIES AMONG HEBREW SCHOLARS RELATING TO RITUALISTIC INKS—THE CLASS OF INKS EMPLOYED BY THE FRENCH AND GERMAN JEWS—CONVENTION OF REPRESENTATIVES FROM JEWISH CENTERS—SUBMISSION OF THEIR DIFFERENCES TO MAIMONIDES—HE DEFINES TALMUDIC INK—SIXTH CENTURY REFERENCE TO "GALL" INK—ASSERTION OF HOTZ-OSTERWALD THAT EXCLUSIVE OF THE INDIAN INK, THE WRITING PIGMENTS OF ANTIQ-

UITY HAVE NEVER BEEN INVESTIGATED—HIS BELIEF THAT YEAST FORMED A PORTION OF THEM—SOME OTHER OBSERVATIONS ON THIS SUBJECT—ANCIENT FORMULAS ABOUT THE LEES OF WINE IN INK-MAKING—COMMENTS ON INK-MAKING BY PLINY—ANCIENT FORMULA OF POMEGRANATE INK— SECRETA BY THE MONK THEOPHILUS—WHAT THE, THORN TREE HE REFERS TO REALLY IS—IDENTITY OF THE MYROBOLAM INK OF THE MOST REMOTE ANTIQUITY WITH THE POMEGRANATE INK OF THE MIDDLE AGES— THE USES OF THE ACACIA TREE.

MOST of the documents of early mediaeval times which remain to us containing ink in fairly good condition, like charters, protocols, bulls, wills, diplomas, and the like, were written or engrossed with "Indian" ink, in which respect we of the present century continue to follow such established precedent when preparing important written instruments. It is not remarkable, therefore, that the black inks of the seventh, eighth, ninth and tenth centuries preserve their blackness so much better than many belonging to succeeding ages, including a new class of inks which could not stand the test of time.

During the twelfth and first years of the thirteenth centuries there were bitter controversies among Talmudic (Hebrew) scholars, relative to the character of the ink to be employed in the preparation of ritualistic writings. Nice distinctions were drawn as to the real meaning of the word deyo as understood by the Jews of the western part of the world, and the Arabic word alchiber, as then understood nearer Palestine and the other eastern countries.

The French Jews were using "tusche" (typical of the "Indian" ink), while the Germans were employing "pomegranate" and "gall" inks. Representatives from interested religious Jewish centers came together and resolved to submit their differences for final adjustment to Maimonides, born in Spain, A. D. 1130 , and died A. D. 1204—the then greatest living Hebrew theologian and authority on biblical and rabbinical laws. Discarding all side issues, their differences were seemingly incorporated into three questions and thus propounded to him:

1. Is the Talmudic deyo identical with alchiber?

2. Of what ingredient should the Talmudic deyo consist, if it is not the same as alchiber?

3. Is alchiber to be understood as relating to the gall-apple and chalkanthum (blue vitriol)?

To the first and third questions Maimonides declared that deyo and alchiber were not identical; and for the reasons that the Talmud declares deyo to be a writing material which does not remain on the surface on which it is placed and to be easily effaced. On the other hand alchiber contains gum and other things which causes it to adhere to the writing surface.

To the second question he affirmed that the Talmud distinguishes a double kind of deyo, one containing little or no gum and being a fluid, and the other referring to "pulverized coal of the vine, soot from burning olive oil, tar, rosin and honey, pressed into plates to be dissolved in water when wanted for use." Furthermore, while the Talmud excludes the use of certain inks of which iron vitriol was one, it does not exclude atramentum, (chalkanthum, copper vitriol), because the Talmud never speaks of it. He insisted that the Talmud requires a dry ink (deyo).

As one of the last entries made in the Talmud (a great collection of legal decisions by the ancient Rabbis, Hebrew traditions, etc., and believed to have been commenced in the second century of the Christian era) is claimed to belong to the sixth century, mentions gall-apples and iron (copper) vitriol, it must have referred to "gall" ink. Further investigation discloses the fact that such galls were of Chinese origin and as we know they do not contain the necessary ferment which the aleppo and other galls possess for inducing a transformation of the tannin into gallic acid, no complete union could therefore obtain. Hence the value of this composition was limited until the time when yeast and other materials were introduced to overcome its deficiencies.

Hotz-Osterwald of Zurich, antiquarian and scholar, has asserted that with the exception of the carbon inks employed on papyrus, the writing pigments of antiquity and the Middle Ages have scarcely been investigated. The dark to light-brown pigment, hitherto a problem, universally used on parchment, he contends upon historical, chemical and microscopic evidence is identical with oenocyanin and was prepared for the most part from yeast, and was first employed as a pigment. Contrary to the general opinion it contains no iron, except frequently accidental traces, and after its appearance in Greece in the third century, it formed almost exclusively the ink of the ancient manuscripts, until displaced by the gallate inks, said to have been introduced by the Arabians. These accidental traces of iron were due to the employment of iron vessels in the making of the ink.

My own observations in this direction confirm and establish the fact that it was the custom in the early centuries of the Christian era to utilize yeast or an analogous compound as part of the composition of ink, to which was added sepia, or the rind of the pomegranate apple previously dissolved by heat in alkaline solutions.

This analogous compound was probably the material procured from wine lees (dregs), deposited after fermentation has commenced, and which after considerable application of heat yields not only most of the tannin contained in the stones and fruit stalks, but a viscid compound characteristic of gelatine and of a red-purple color which in course of time changes to brown.

Bloxam says that the coloring matter of grapes and of red wine appears to be "cyanin."

One of the methods of treating wine lees, as translated in the eighteenth century from an old Italian secreta, is sufficiently curious to partly quote:

"Dry the Lees (dregs) of wine with a gentle fire and fill with them two third of a large earthen Retort, place this retort in a reverberatory furnace, and fitting it to a large receiver, give a small fire to it to heat the Retort by degrees, and drive forth an insipid phlegm; when vapours begin to rise, you must take out the phlegm and luting carefully the junctures of your vessels, quicken the fire little by little until you find the receiver filled with white clouds; continue it in this condition, and you perceive the receiver to cool, raise the fire to the utmost extremity, and continue it so, until there arise no more vapours. When the vessels are cold unlute the receiver, and shaking it to make the Volatile salt, which sticks to it, fall to the bottom, pour it all into a bolt-head; fit it to a Head with a small receiver; lute well the junctures and placing it in sand, give a little fire under it, and the volatile salt will rise and stick to the head, and the top of the Bolt-head; take off your head and set on another in its place; gather your salt and stop it tip quickly, for it easily dissolves into a liquor; continue the fire, and take care to gather the Salt according as you see it appear; but when there rises no more salt, a liquor will distill, of which you must draw about three ounces, and put out the fire," &c.

The "lees of wine," in connection with the ancient methods of ink-making is also referred to by the younger Pliny in his twenty-fifth book, which the Edinburgh Review has carefully translated and printed:

"INK (or literally) BLACKING.—Ink also may be set down among the artificial (or compound) drugs, although it is a mineral derived from two sources. For, it is sometimes developed in the form of a saline efflorescence,—or is a real mineral of sulphureous color—chosen for this purpose. There have been painters who dug up from graves colored coals (CARBON). But all these are useless and new-fangled notions. For it is made from soot in various forms, as (for instance) of burnt rosin or pitch. For this purpose, they have built manufactories not emitting that smoke. The ink of the very best quality is made from the smoke of torches. An inferior article is made from the soot of furnaces and bath-house chimneys. There are some (manufacturers) also, who employ the dried lees of wine; and they do say that if the lees so employed were from good wine, the quality of the ink is thereby much improved. Polygnotus and Micon, celebrated painters at Athens, made their black paint from burnt grape-vines; they gave it the name of TRYGYNON. APELLES, we are told, made HIS from burnt ivory, and called it elephantina 'ivory-black.' Indigo has been recently imported,— a substance whose composition I have not yet investigated. The dyers make theirs from the dark crust that gradually accumulates on brass-kettles. Ink is made also from torches (pine-knots), and from char-coal pounded fine in mortars. 'The cuttlefish' has a remarkable qual-ify in this respect; but the coloring-matter which it produces is not used in the manufacture of ink. All ink is improved by exposure to the sun's rays. Book-writers' ink has gum mixed with it,—weavers' ink is made up with glue. Ink whose materials have been liquified by the agency of an acid is erased with great difficulty."

There are but few exceptions respecting the general sameness of ink receipts of the succeeding centuries, one of which is the "Pomegranate," credited to the seventh century but really belonging to an earlier period:

"Of the dried Pommegranite (apple) rind take an ounce, boil it in a pint of water until 3/4 be gone; add 1/2 pint of small beer wort and once more boil it away so that only a 1/4 pint remain. After you shall have strained it, boiling hot through a linnen cloth and it comes cold, being then of a glutinous consistence, drop in a 'bit' of Sal Alkali and add as much warm water as will bring it to a due fluidity and a gold brown color for writing with a pen."

Following this formula and without any modifications, I obtained an excellent ink of durable quality, but of poor color, from a standpoint of blackness.

A less ancient "Secreta," signed by the Italian monk "Theophilus," who lived about the commencement of the eleventh century, is most interesting:

"To make ink, cut for yourself wood of the thorn-trees in April or May, before they produce flowers or leaves, and collecting them in small bundles, allow them to lie in the shade for two, three, or four weeks, until they are somewhat dry. Then have wooden mallets, with which you beat these thorns upon another piece of hard wood, until you peel off the bark everywhere, put which immediately into a barrelful of water. When you have filled two, or three, or four, or five barrels with bark and water, allow them so to stand for eight days, until the waters imbibe all the sap of the bark. Afterwards put this water into a very clean pan, or into a cauldron, and fire being placed under it, boil it; from time to time, also, throw into the pan some of this bark, so that whatever sap may remain in it may be boiled out. When you have cooked it a little, throw it out, and again put in more; which done, boil down the remaining water unto a third part, and then pouring it out of this pan, put it into one smaller, and cook it until it grows black and begins to thicken; add one third part of pure wine, and putting it into two or three new pots, cook it until you see a sort of skin show itself on the surface; then taking these pots from the fire, place them in the sun until the black ink purifies itself from the red dregs. Afterwards take small bags of parchment carefully sewn, and bladders, and pouring in the pure ink, suspend them in the sun until all is quite dry; And when dry, take from it as much as you wish, and temper it with wine over the fire, and, adding a little vitriol, write. But, if it should happen through negligence that your ink be not black enough, take a fragment of the thickness of a finger and putting it into the fire, allow it to glow, and throw it directly into the ink."

After reciting many receipts which pertain to other arts, this good old monk concludes:

"When you shall have re-read this often, and have committed it to your tenacious memory, you shall thus recompense me for this care of instruction, that, as often as you shall successfully have made use of my work, you pray for me for the pity of omnipotent God, who knows that I have written these things which are here arranged, neither through love of human approbation, nor through desire of temporal reward, nor have I stolen anything precious or rare through envious jealousy, nor have I kept back anything reserved for myself alone; but, in augmentation of the honour and glory of His name, I have consulted the progress and hastened to aid the necessities of many men."

The "thorn" trees which Theophilus mentions are asserted by some writers (with whom I do not agree) to be those commonly known as the "Norway spruce," a species of pine of lofty proportions sometimes rising to the height of 150 feet with a trunk from four to five feet in diameter. It lives to a great age believed to exceed in many instances 450 years. The leaves (needles, thorns) are short but stand thickly upon the branches and are of a dusky green color shining on the upper surface; the fruit is nearly cylindrical in form and of a purple color covered with scales ragged at the edges. It is a native of Europe and Northern Asia. It furnishes the material known as Burgundy pitch which is obtained by removing the juice which is secreted in the bark of the tree; it is purified by a melting process and straining either through a cloth or a layer of straw. It gives forth a peculiar odor not unpleasant, resembling turpentine.

The Burgundy pitch or rosin is soluble in hot alcohol (spirits of wine).

An ink prepared after the method laid down by this monk, assuming that he referred to the spruce-pine, while troublesome to write with, would be almost as lasting as "Indian" ink and would be most difficult to erase from parchment into which it would be absorbed due to its alcoholic qualities.

"The ink," remarks Montfaucon, "which we see in the most ancient Greek manuscripts, has evidently lost much of its pristine blackness; yet neither has it become altogether yellow or faint, but is rather tawny or deep red, and often not far from a vermillion." While there are some monuments of this kind of ink in fair condition of the fourth and succeeding centuries, they aggregate but a very small proportion of the vast number of principally Indian ink specimens which remain to us of those epochs. As exemplars, however, of a forgotten class of inks belonging to a still more remote antiquity, careful research adduces certain proof of their existence more than nine hundred years before the Christian era commenced.

Reference has earlier been made to the ancient Myrobolam ink, which was characteristically the same in color phenomena as those which Montfaucon mentions. These "tawny" colored inks I estimate were products obtained from the "thorn" trees spoken of by the monk Theophilus. The thorn trees were of two species. The pomegranate, anciently called the "Punic apple," because it was largely employed by the Carthagenians for the purposes of dyeing and tanning; and the acacia, known in Egyptian times as the lotus. The former was held in such high esteem that the Arabians and Egyptians made it an emblem to designate one of their dieties and termed it raman.

The products of these thorn, trees were collectively used together as ink, most of the tannin being obtained from the pomegranate, and the gum from the acacia.

CHAPTER VIII.

MEDIAEVAL INK.

INK SECRETAS OF THE TWELFTH CENTURY COMPARED WITH EARLIER ONES—APPEARANCE OF TANNO-GALLATE OF IRON INK IN THE TWELFTH CENTURY—ITS INTRODUCTION LOCATES THE EPOCH WHEN THE MODERN INK OF TO-DAY FIRST CAME INTO VOGUE—ITS APPROVAL AND ADOPTION BY THE FATHERS OF THE CHURCH—THE INVENTION NOT ITALIAN BUT ASIATIC—ITS ARRIVAL FROM ASIA FROM THE WEST AND NOT THE EAST—APPEARANCE ABOUT THE SAME TIME OF LINEN OR MODERN PAPER—SETTLEMENT OF OLD CONTROVERSIES ABOUT ANCIENT SO-CALLED COTTON PAPER-DE VINNE'S COMMENT ABOUT PAPER AND PAPER-MAKING—CURIOUS CONTRACT OF THE FOURTEENTH CENTURY.

THE "Secretas" of the twelfth century, in so far as they relate to methods of making ink, indicate many departures from those contained in the more ancient ones. Frequent mention is made of sour galls, aleppo galls, green and blue vitriol, the lees of wine, black amber, sugar, fish-glue and a host of unimportant materials as being employed in the admixture of black inks. Combinations of some of these materials are expressed in formulas, the most important one of which details with great particularity the commingling together of an infusion of nut-galls, green vitriol (sulphate of iron) and fish-glue (isinglass); the two first (tanno-gallate of iron) when used alone, forms the sole base of all unadulterated "gall" inks.

Dates are appended to some of these ink and other formulas. The "tanno-gallate of iron" one has, however, no date. But as it appears closely following a date of A. D. 1126, it must have been written about that time.

Documents, public and private, bearing dates nearly contemporary with that era, written in ink of like type, are still extant, confirming in a remarkable degree the "Secreta" formula, and establishing the fact that the first half of the twelfth century marks the epoch in which the "gall" or modern ink of today came into vogue.

Its adoption by the priests stamped it with the seal of the Church and the arrival from the West about the same period of flax or linen paper with the added fact that these assimilated so well together, later placed them both on the popular basis which has continued to the present time.

While the Secreta which contains the "gall" ink formula is of Italian origin, the invention of this ink belongs solely to an Asiatic country, from whence in gradual stages by way of Arabia, Spain and France, it finally reached Rome. Thence, through the Church, information about it was conveyed to wherever civilization existed.

We are not confined in our investigations of ancient MSS. to any particular locality or date, as the twelfth, thirteenth, fourteenth and fifteenth centuries are prolific of "gall" ink monuments covering an immense territory. Such inks when used unadulterated, remain in an almost pristine color condition; while the other inks to which some pigment or color had been added, probably to make them more agreeable in appearance and more free-flowing, with a mistaken idea of improving them, are much discolored and in every instance present but slight indications of their original condition.

The question of the character of the paper employed during these eras, composed of different kinds of fibrous vegetable substances, possesses some importance when discussing its relationship to inks. Many authors certify to the manufacture and use of "cotton" in the eleventh, twelfth and later centuries. Madan, however, in treating this subject, makes the following comments which are in line with my own observations:

"Paper has for long been the common substance for miscellaneous purposes of ordinary writing, and has at all times been formed exclusively from rags (chiefly of linen) reduced to pull), poured out on a frame in a thin watery sheet, and gradually dried and given consistence by the action of heat. It has been a popular belief, found in every book till 1886 (now entirely disproved, but probably destined to die hard), that the common yellowish thick paper, with rough fibrous edge, found especially in Greek MSS. till the fifteenth century, was paper of quite another sort, and made of cotton (charta bombycna, bombyx being usually silk, but also used of any fine fibre such as cotton). The microscope has at last conclusively shown that these two papers are simply two different kinds of ordinary linen-rag paper."

De Vinne speaking, of paper and paper-making says:

"The gradual development of paper-making in Europe is but imperfectly presented through these fragmentary facts. Paper may have been made for many years before it found chroniclers who thought the manufacture worthy of notice. The Spanish paper-mills of Toledo which were at work in the year 1085, and an ancient family of paper-makers which was honored with marked favor by the king of Sicily in the year 1102, are carelessly mentioned by contemporary writers as if paper-making was an old and established business. It does not appear that paper was a novelty at a much earlier period. The bulls of the popes of the eighth and ninth centuries were written on cotton card or cotton paper, but no writer called attention to this card, or described it as a new material. It has been supposed that this paper was made in Asia, but it could have been made in Europe. A paper-like fabric, made from the barks of trees, was used for writing by the Longobards in the seventh century, and a coarse imitation of the Egyptian papyrus, in the form of a strong brown paper, had been made by the Romans as early as the third century. The art of compacting in a web the macerated fibres of plants seems to have been known and practised to some extent in Southern Europe long before the establishment of Moorish paper-mills.

"The Moors brought to Spain and Sicily not an entirely new invention, but an improved method of making paper, and what was more important, a culture and civilization that kept this method in constant exercise. It was chiefly for the lack of ability and lack of disposition to put paper to proper use that the earlier European knowledge of paper- making was so barren of results. The art of book- making as it was then practised was made subservient to the spirit of luxury more than to the desire for knowledge. Vellum was regarded by the copyist as the only substance fit for writing on, even when it was so scarce that it could be used only for the most expensive books. The card-like cotton paper once made by the Saracens was certainly known in Europe for many years before its utility was recognized. Hallam says that the use of this cotton paper was by no means general or frequent, except in Spain or Italy, and perhaps in the south of France, until the end of the fourteenth century. Nor was it much used in Italy for books.

"Paper came before its time and had to wait for recognition. It was sorely needed. The Egyptian manufacture of papyrus, which was in a state of decay in the seventh century, ceased entirely in the ninth or tenth. Not many books were written during this period, but there was then, and for at least three centuries afterwards, an unsatisfied demand for something to write upon. Parchment was so scarce that reckless copyists frequently resorted to the desperate expedient of effacing the writing on old and lightly esteemed manuscripts. It was not a difficult task. The writing ink then used was usually made of lamp-black, gum and vinegar; it it had but a feeble encaustic property, and it did not bite in or penetrate the parchment. The work of effacing this ink was accomplished by moistening the parchment with a weak alkaline solution and by rubbing it with pumice stone. This treatment did not entirely obliterate the writing, but made it so indistinct that the parchment could be written over the second time. Manuscripts so treated are now known as palimpsests. All the large European public libraries have copies of palimpsests, which are melancholy illustrations of the literary tastes of many writers or bookmakers during the Middle Ages. More convincingly than by argument they show the utility of paper. Manuscripts of the Gospels, of the Iliad, and of works of the highest merit, often of great beauty and accuracy, are dimly seen underneath stupid sermons, and theological writings of a nature so paltry that no man living cares to read them. In Some instances the first writing has been so thoroughly scrubbed out that its meaning is irretrievably lost.

"Much as paper was needed, it was not at all popular with copyists; their prejudice was not altogether unreasonable, for it was thick, coarse, knotty, and in every way unfitted for the display or ornamental penmanship or illumination. The cheaper quality, then known as cotton paper, was especially objectionable. It seems to have been so badly made as to need governmental interference. Frederick II, of Germany, in the year 1221, foreseeing evils that might arise from bad paper, made a decree by which he made invalid all public documents that should be put on cotton paper, and ordered them within two years to be transcribed upon parchment. Peter II, of Spain, in the year 1338, publicly commanded the paper-makers of Valencia and Xativa to make their paper of a better quality and equal to that of an earlier period.

"The better quality of paper, now known as linen paper, had the merits of strength, flexibility, and durability in a high degree, but it was set aside by the copyists because the fabric was too thick and the surface was too rough. The art of calendering or polishing papers until they were of a smooth, glossy surface, which was then practised by the Persians, was unknown to, or at least unpractised by, the early European makers. The changes or fashion in the selection of writing papers are worthy of passing notice. The rough hand-made papers so heartily despised by the copyists of the thirteenth century are now preferred by neat penmen and skilled draughtsmen. The imitations of mediaeval paper, thick, harsh, and dingy, and showing the marks of the wires upon which the fabric was couched, are preferred by men of letters for books and for correspondence, while highly polished modern plate papers, with surfaces much more glossy than any preparation of vellum, are now rejected by them as finical and effeminate.

"There is a popular notion that the so-called inventions of paper and xylographic printing were gladly welcomed by men of letters, and that the new fabric and the new art were immediately pressed into service. The facts about to be presented in succeeding chapters will lead to a different conclusion. We shall see that the makers of playing cards and of image prints were the men who first made extended use of printing, and that self-taught and unprofessional copyists were the men who gave encouragement to the manufacture of paper. The more liberal use of paper at the beginning of the fifteenth century by this newly- created class of readers and book-buyers marks the period of transition and of mental and mechanical development for which the crude arts of paper- making and of black printing had been waiting for centuries. We shall also see that if paper had been ever so cheap and common during the Middle Ages, it would have worked no changes in education or literature; it could not have been used by the people, for they were too illiterate; it would not have been used by the professional copyists, for they preferred vellum and despised the substitute.

"The scarcity of vellum in one century, and its abundance in another, are indicated by the size of written papers during the same periods. Before the sixth century, legal documents were generally written upon one side only; in the tenth century the practice of writing upon both sides of the vellum became common. During the thirteenth century valuable documents were often written upon strips two inches wide and but three and a half inches long. At the end of the fourteenth century these strips went out of fashion. The more general use of paper had diminished the demand for vellum and increased the supply. In the fifteenth century, legal documents on rolls of sewed vellum twenty feet in length were not uncommon. All the valuable books of the fourteenth century were written on vellum. In the library of the Louvre the manuscripts on paper, compared to those on vellum, were as one to twenty-eight; in the library of the Dukes of Burgundy, one-fifth of the books were of paper. The increase in the proportion of paper books is a fair indication of the increasing popularity of paper; but it is obvious that vellum was even then considered as the more suitable substance for a book of value."

The curious contract belonging to the fourteenth century which follows, is a literal copy of the original. It does not seem to specify whether the book is to be made of vellum or paper. In other respects the minute details no doubt prevented any misunderstanding between the contracting parties.

"August 26th, 1346—There appeared Robert Brekeling, scribe, and swore that he would observe the contract made between him and Sir John Forber, viz., that the said Robert would write one Psalter with the Kalender for the work of the said Sir John for 5 s. and 6 d.; and in the same Psalter, in the same character, a Placebo and a Dirige, with a Hymnal and Collectary, for 4 s. and 3 d. And the said Robert will illuminate ('luminabet') all the Psalms with great gilded letters laid in with colours; and all the large letters of the Hymnal and Collectary will he illuminate with gold and vermillion, except the great letters of double feasts, which shall be as the large gilt letters are in the Psalter. And all the letters at the commencement of the verses shall be illuminated with good azure and vermillion; and all the letters at the beginning of the Nocturns shall be great uncial (unciales) letters, containing V. lines, but the Beatus Vir and Dixit

Dominus shall contain VI. or VII. lines; and for the aforesaid illumination and for colours he [John] will give 5 s. 6 d., and for gold he will give 18 d., and 2 s. for a cloak and fur trimming. Item one robe—one coverlet, one sheet, and one pillow."

CHAPTER IX.

END OF MEDIAEVAL INK.

THE SECRETAS PRECEDE ALCHEMY AND CHEMISTRY—EFFORT TO IMPROVE GALL INKS—VARIATIONS IN INK COLORS—THE USE OF RED INK IN THE NINTH AND TENTH CENTURIES—COLOR COMPARISONS BETWEEN INK WRITINGS OF ITALY, GERMANY, FRANCE, ENGLAND AND SPAIN—HOW TO DETERMINE THE ANTIQUITY OF MSS.—PRACTICES WHICH OBTAINED IN MONASTIC LIBRARIES OF VARIOUS COUNTRIES—-KINDS OF INK EMPLOYED IN LITURGICAL WRITINGS—THE PUBLIC SCRIBES AND THEIR EMPLOYMENTS—EFFORTS TO COUNTERFEIT OLD SCRIPT IN EARLY PRINTED BOOKS—WHEN THEY WERE ABANDONED.

IT is well known that alchemy preceded chemistry and hence the Secreta came first. When the formula for making a real "gall" ink had ceased to be a secret, chemistry was then but little understood. It is not a matter for wonder, therefore, to learn that "gall" ink of the first half of the twelfth century was low in grade and poor in quality. It was a muddy fluid easily precipitated and it deteriorated quickly. A century or more of experimenting was needed to modify or overcome defects, as well as to gain information about the chemical value of the different tannins, the relative proportions of each constituent and the correct methods in its admixture.

There is no written account of this ink being manufactured as an industry until over three hundred years later. Hence, as it appears so frequently of varying degrees of color on documents of the intervening centuries, we are compelled to assume that it was compounded by individuals who had neither chemical knowledge, nor who had made a study or a business of ink-making. Notwithstanding which, its progress seems to have been comparatively rapid and like the same ink of the present day was to be obtained of any quality or kind, whether unadulterated or containing some added color.

Intense black or a black tinged with red-brown characterizes the color of the inks found on the very earliest MSS. Their lasting color phenomena, due to the employment of lampblack and kindred substances even after a lapse of so many ages, is at this late day of no particular moment as they but prove the virtues of the different types of "Indian" inks.

A different set of facts are evident in the inks of mediaeval times which are found to greatly vary according to their ages and locality. But few black inks of the ninth and tenth centuries remain to us. In the MSS. of those centuries a red ink was the prevailing one even to the extent of entire volumes being written with it. In Italy and many other portions of Southern Europe specimens now extant, when compared with those belonging to Germany and other more northern countries, are seen to be blacker and this is also true when those of France and England are compared, the blacker inks belonging to France. With the gradual disappearance of the so-called "Dark Ages," the ink found on Spanish written MSS. of the fourteenth and fifteenth centuries, are notedly of intense blackness while those of some of the other countries appear of a rather faded gray color, and in the sixteenth century, this gray color effect prevailed all over the Christian world.

To revert again to the ink phenomena of the fourteenth and fifteenth centuries which are of Italian origin. In no section of that country or of Europe during those centuries do ink creations possess, in so marked a degree, the variety of color qualities that are seen on those of the city of Florence. Indeed it may be truly said that during those periods more ink written MSS. were produced in that place than all the rest of Europe. These productions of MSS. were not confined to simple ink writings. The heads of religious orders and rulers of the country liked to have artists near them to illuminate their missals and sacred books, besides the decorating of walls in their churches and palaces.

Through this art of illuminating and the painting of miniatures in MSS. books, "oil" painting took root and the day for mere symbols and hieroglyphics was over.

In that city of scholars and wealth it was a fashion and later the custom to acquire Greek, Latin and Oriental MSS. and copy them for circulation and sale. The prices offered were sufficient to stimulate the search and zeal for them. We learn that in the year 1400 "on the square of the Duoma a spacciatore was established whose business was to sell manuscripts often full of mistakes and blunders." Nicholas V, before he became Pope, was nicknamed "Tommaso the Copyist." He is said to have presented to the Vatican library as a gift five thousand volumes of his own creation.

The information of these increasing demands for ancient documents of any kind spread over Europe and portions of Asia, bringing into Florence a great quantity of them, as well as many scholars and copyists. Shiploads of the works of the Byzantine historians arrived from the Golden Horn, and the city became a vast manufactory for duplicating or forging ancient MSS. Parchment and vellum were too costly to employ very much, so most of them were of paper. Vespaciano, one of the many engaged in this business and who lived in 1464, found it necessary in order to reduce the cost of production, to become a paper merchant. In writing to a friend he says:

"I engaged forty-five copyists and in twenty- two months had completed two hundred volumes, which included some Greek and Latin as well as many Oriental writings."

The reading and judging of manuscripts are now known as the science of diplomatics. To determine their antiquity or genuineness requires the nicest distinctions and care, irrespective of alleged dates (whether exhibited by Roman numbers or the Arabic one which we continue to employ, and which first made their appearance near the commencement of the twelfth century). The inks as already mentioned and used on them, as we shall see, serve fully as much in estimating authenticity or genuineness as does combined together,—the style of the writing, the miniatures, vignettes and arabesques (if any), the colors, covers, materials, ornamentation and the character of their contents.

With the re-establishment of learning in the fifteenth century and the creation of alleged stable governments, who may perhaps have realized the necessity for an ink of enduring good commercial and record qualities, so-called "gall" inks were chosen as best possessing them, and were made and employed with varying results even more than the ancient "Indian" inks.

Mediaeval practices in relation to ink and other writing materials as well as the monastic libraries of which England, France, Germany and Italy possessed many during the thirteenth, fourteenth, and more particularly the fifteenth centuries, were governed by established rules.

The libraries of such institutions were placed by the abbot under the sole charge of the "armarian," an officer who was made responsible for the preservation of the volumes under his care; be was expected frequently to examine them, lest damp or insects should injure them; he was to cover them with wooden covers to preserve them and carefully to mend and restore any damage which time or accident might cause; he was to make a note of any book borrowed from the library, with the name of the borrower; but this last rule applied only to the less valuable portion of it, as the "great and precious books" could only be lent by the permission of the abbot himself. It was also the duty of the armarian to have all the books in his charge marked with their correct titles, and to keep a perfect list of the whole. Some of these catalogues are still in existence and are curious and interesting in their exemplification of the kinds of ink employed and as indicative of the state of literature in the Middle Ages, besides presenting the names of many authors whose works have never reached us. It was also the duty of the armarian, under the orders of his superior, to provide the transcribers of manuscripts with the writings which they were to copy, as well as all the materials necessary for their labors, to make bargains as to payment, and to superintend the work during their progress.

These transcribers, Mr. Maitland in his "Dark Ages" tells us, were monks and their clerks, some of whom were so skilled that they could perform all the different branches. They were exhorted by the rules of their order to learn writing, and to persevere in the work of copying manuscripts as being one most acceptable to God; those who could not write were recommended to bind books. This was in line with the behest of the famous monk Alciun who lived in the eighth century and who entreated all to employ themselves in copying books, saying:

"It is a most meritorious work, more useful to the health than working in the fields, which profits only a man's body, while the labour of a copyist profits his soul."

When black ink was used in liturgical writings, the title page and heads of chapters were written in red ink; whence comes the term rubric. Green, purple, blue and yellow inks were sometimes used for words, but chiefly for ornamenting capital letters.

A large room was in most monasteries set apart for such labors and here the general transcribers pursued their avocations; in addition, small rooms or cells, known also as scriptoria, occupied by such monks as were considered, from their piety and learning, to be entitled to the indulgence, and used by them for their private devotions, as well as for the purpose of transcribing works for the use of the church or library. The scriptoria were frequently enriched by donations and bequests from those who knew the value of the works carried on in them, and large estates were often devoted to their support.

"Meanwhile along the cloister's painted side,
The monks—each bending low upon his book
With head on hand reclined—their studies plied;
Forbid to parley, or in front to look,

Lengthways their regulated seats they took:
The strutting prior gazed with pompous mien,
And wakeful tongue, prepared with prompt rebuke,
If monk asleep in sheltering hood was seen;
He wary often peeped beneath that russet screen.

"Hard by, against the window's adverse light,
Where desks were wont in length of row to stand,
The gowned artificers inclined to write;
The pen of silver glistened in the hand
Some of their fingers rhyming Latin scanned;

Some textile gold from halls unwinding drew,
And on strained velvet stately portraits planned;
Here arms, there faces shown in embryo view,
At last to glittering life the total figures grew."
—FOSBROOKE.

The public scribes of those days were employed mostly by secular individuals, although subject to be called upon at any moment by the fathers of the church. They worked in their homes except when any valuable work was to be copied, then in that of their employer, who boarded and lodged them during the time of their engagement.

To differentiate the character of the class of pigments or materials then employed in making colored inks, from those of the more ancient times is difficult; because we not only find many of like character but of larger variety. These were used more for purposes of illuminating and embellishing than for regular writing.

Even when printing had been invented spaces were frequently left, both in the block books and in the earliest movable type, for the illumination by hand, of initial letters so as to deceive purchasers into the belief that the printed type which was patterned closely after the forms of letters employed in MSS. writings was the real thing. The learned soon discovered such frauds and thereafter these practices were abandoned.

CHAPTER X.

RENAISSANCE INK.

INK OF GRAY COLOR BELONGING TO THE SIXTEENTH CENTURY AND ITS CAUSES—INFLUENCE OF THE FATHERS OF THE CHURCH RESPECTING INK DURING THE DARK AGES—THE REFORMATION AND HOW IT AFFECTED MEDIAEVAL MSS.—REMARKS OF BALE ABOUT THEIR DESTRUCTION— QUAINT INK RECEIPT OF 1602—SELECTION

FROM THE TWELFTH NIGHT RELATING TO PEN AND INK—GENERAL CONDITIONS WHICH OBTAINED UNTIL 1626—THE FRENCH GOVERNMENT AWARDS AN INK CONTRACT IN THAT YEAR—OTHER GOVERNMENTS ADOPT THE FRENCH FORMULA—INKS OF THE SEVENTEENTH CENTURY ALMOST PERFECT IN THEIR COLOR PHENOMENA— NO ADDED COLOR EMPLOYED IN THEIR MANUFACTURE.

THE gray color of most of the inks found on documents written in the sixteenth century is a noteworthy fact. Whence its cause is a matter for considerable speculation. The majority of these inks unquestionably belong to the "gall" class and if prepared after the formulas utilized in preceding centuries should indicate like color phenomena. As these same peculiarities exist on both paper, vellum and parchment, it cannot be attributed to their use. Investigations in many instances of the writings indicate the exercise of a more rapid pen movement and a consequent employment of inks of greater fluidity than those of an earlier history. Such fluidity could only be obtained by a reduction of the quantity of gummy vehicles together with an increase of ink acidity. The acids which had theretofore been more or less introduced into inks, except oxalic acid, could not effect such results. Consequently, as the monuments of this gray ink phenomena are to be found belonging to all the portions of the Christian world, with a uniformity that is certainly remarkable, it becomes a fair deduction to assume that the making of inks bad passed into the hands of regular manufacturers who adulterated them with "added" color.

We can well believe that the influences which the fathers of the Church exerted during the thousand years known as the "Dark Ages," in respect to ink and kindred subjects, must have been very great. That they endeavored to perpetuate for the benefit of succeeding generations in book and other forms, this kind of information, which they distributed throughout the world we know to be true. Most of these sources of ink information, however, gradually disappeared as constituting a series of sad events in the unhappy war which followed their preparation.

The Reformation began in Germany in the first quarter of the sixteenth century, and with it the eighty years of continual religious warfare which followed. During this period the priceless MSS.

books of information, historical, literary and otherwise, contained in the monastic libraries outside of Italy were burnt.

We are told:

"In England cupidity and intolerance destroyed recklessly. Thus, after the dissolution of monastic establishments, persons were appointed to search out all missals, books of legends, and such 'superstitious books' and to destroy or sell them for waste paper; reserving only their bindings, when, as was frequently the case, they were ornamented with massive gold and silver, curiously chased, and often further enriched with precious stones; and so industriously had these men done their work, destroying all books in which they considered popish tendencies to be shown by illumination, the use of red letters, or of the Cross, or even by the—to them —mysterious diagrams of mathematical problems— that when, some years later, Leland was appointed to examine the monastic libraries, with a view to the preservation of what was valuable in them, he found that those who had preceded him had left little to reward his search."

Bale, himself an advocate for the dissolution of monasteries, says:

"Never had we bene offended for the losse of our lybraryes beyng so many in nombre and in so desolute places for the moste parte, yf the chief monuments and moste notable workes of our excellent wryters had bene reserved, yf there had bene in every shyre of Englande but one solemyne lybrary for the preservacyon of those noble workes, and preferrments of good learnyuges in our posteryte it had bene yet somewhat. But to destroye all without consyderacyon is and wyll be unto Englande for ever a most horryble infamy amonge the grave senyours of other natyons. A grete nombre of them wych purchased of those superstycyose mansyons reserved of those lybrarye bokes, some to serve theyr jaks, some to scoure theyr candelstyckes, and some to rubb theyr bootes . some they solde to the grossers and sope sellers, and some they sent over see to the bokebynders, not in small nombre, but at tymes whole shippesful. I knowa merchant man, whyche shall at thys tyme be namelesse, that boughte the content-, of two noble lybraryes for xl shyllyngs pryce, a shame it is to be spoken. Thys stuffe hathe he occupyed in the stide of greve paper for the space of more than these ten years, and yet hathe store ynough for as many years to come. A prodyguous example is thys, and to be abhorred of all men who love theyr n atyon as they shoulde do."

Passing to later epochs, A. D. 1602, the following quaint receipt proves interesting as showing that the "gall" inks were well known at that time:

"To make common Ink, of Wine take a quart,
Two ounces of Gumme, let that be a part;
Five ounces of Galls, of Cop'res take three,
Long standing doth make it the better to be;
If Wine ye do want, raine water is best,
And then as much stuffe as above at the least,
If the Ink be too thick, put Vinegar in,
For water doth make the colour more dimme."

Shakespeare in his Twelfth Night III, 2, has also referred to them in the following amusing strain:

"Go write it in a martial hand; be curst and brief; it is no matter how witty, so it be eloquent, and full of invention; taunt him with the license of ink; if thou thou'st him thrice, it shall nor be amiss; and as many lies as will lie on a sheet of paper, although the sheet were big enough for the bed of Ware in England, set 'em down; go, about it. Let there be gall enough in thy ink, though thou write with a goose pen, no matter: about it."

The general black ink conditions for a period of at least three hundred years, if we exclude the sixteenth century, had been but repetitions of each other. They so remained until the year 1626, when the French government concluded an arrangement with a chemist by the name of Guyot, for the manufacture of a "gall" ink WITHOUT added color and which thereby guaranteed and insured more sameness in respect to desirable ink qualities. That government with a few modifications relative to the proportions of ingredients continued its employment, which was followed by the contemporaneous writers. Other governments later partially adopted the French formulas while some of them gave the matter no attention, although their records and those of the cities or towns not only of Europe but early America, the United States and Canada are found in most instances to have been written with an ink of this character.

Where prior to 1850, inks containing a different base (with the single exception of indigo) were used, they have either disappeared or nearly so and it is not an infrequent occurrence among those who are accustomed to examine old records to find that signatures or dates to valuable instruments, pages of writings and indeed sometimes the writings in an entire book are more or less obliterated.

The black inks of a large portion of the seventeenth century, on documents of every kind, are found to be nearly perfect as to color conditions, which is evidence of the extreme care used in their preparation and the exclusion of "added" color in ink manufacture.

CHAPTER XI.

ANCIENT INK TREATISES.

INK TREATISES OF THE FIFTEENTH, SIXTEENTH AND SEVENTEENTH CENTURIES—JOHN BAPTISTA PORTA AUTHOR OF THE FIRST—SECRET INKS—-NERI, CANEPARIUS, BOREL, MERRET, KUNCKEL AND OTHER AUTHORS WHO REFER TO INK MANUFACTURE—PROGRESS OF THE ART OF HANDWRITING ILLUSTRATED IN THE NAMES OF OVER A HUNDRED CALLIGRAPHERS CHRONOLOGICALLY ARRANGED.

THE literature of the fifteenth, sixteenth and seventeenth centuries on the subject of black and colored ink formulas, secret inks, etc., is both diversified and of considerable importance. The following authors and citations are deemed the most noteworthy:

John Baptista Porta, of Naples, born A. D. 1445 and died A. D. 1515, is best known as the inventor of the "camera obscuro;" was also the author of many MSS. books compiled; he says,

"As the results of discussions of long years held at my own house which is known as de Secreti, and into which none can enter unless he claim to be an inventor of new discoveries."

Two of these treatises which were extant in the first half of the seventeenth century, dated respectively 1481 and 1483, dwell at great length on SECRET inks and specifically mention as translated into the English of the time "sowre galls in white wine," and "vitriol;" repeating Italian formulas pertaining to the "Secreta" of the twelfth century.

About secret ink he tells us:

"There are many and almost infinite ways to write things of necessity, that the Characters shall not be seen, unless you dip them into waters, or put them near the fire, or rub them with dust, or smeer them over. * * * * * * * *

"Let Vitriol soak in Boyling water: when it is dissolved, strain it so long till the water grow clear: with that liquor write upon paper: when they are dry they are not seen. Moreover, grinde burnt straw and Vinegar: and what you will write in the spaces between the former lines, describe at large. Then boyl sowre Galls in white Wine, wet a spunge in the liquor: and when you have need, wipe it upon the paper gently, and wet the letters so long until the native black colour disappear, but the former colour, that was not seen, will be made apparent. Now I will show in what liquors paper must be soaked to make letters to be seen. As I said, Dissolve Vitriol in water: then powder Galls finely, and soak them in water: let them stay there twenty-four hours: filtre them through a linen cloth, or something else, that may make the water clear, and make letters upon the paper that you desire to have concealed: send it to your Friend absent: when you would have them appear, dip them in the first liquor, and the letters will presently be seen. * * * * * * * *

If you write with the juice of Citrons, Oranges, Onyons, or almost any sharp things, if you make it hot at the fire, their acrimony is presently discovered: for they are undigested juices, whereas they are detected by the heat of the fire, and then they show forth those colours that they would show if they were ripe. If you write with a sowre Grape that would be black, or with Cervices; when you hold them to the fire they are concocted, and will give the same colour they would in due time give upon the tree, when they were ripe. Juice of Cherries, added to Calamus, will make a green: to sowbread a red: so divers juices of Fruits will show divers colours by the fire. By these means Maids sending and receiving love-letters, escape from those that have charge of them. There is also a kind of Salt called Ammoniac: this powdered and mingled with water, will write white letters, and can hardly be distinguished from the paper, but hold them to the fire, and they will shew black."

With respect to the preparation of black and colored inks and also colors: Antonio Neri, an Italian author and chemist who lived in the sixteenth century, in his treatise seems not only to have laid the foundation for most of the receipts called attention to by later writers during the two hundred years which followed, but to have been the very first to specify a proper "gall" ink and its formula, as the most worthy of notice.

Pietro Caneparius, a physician and writer of Venice, A. D. 1612, in his work De Atrametis, gives a more extensive view about the preparation and composition of inks and adopts all that Neri had given, though he never quotes his name, and adds—"hitherto published by no one." He does however mention many valuable particulars which were omitted by Neri. Most of his receipts are about gold, silver and nondescript inks, with directions for making a great variety for secret writing and defacing. This book revised and enlarged was republished in London, 1660.

In 1653 Peter Borel, who was physician to Louis XIV, King, of France published his "Bibliotheca Chemica," which contains a large number of ink receipts, two of which may be characterized as "iron and gall" ones. They possess value on account of the relative proportions indicated between the two chemicals. The colored ones, including gold, silver and sympathetic inks are mostly repetitions of those of Neri and Caneparius. The French writers, though, speak of his researches in chemistry as "somewhat credulous."

Christopher Merret, an English physician and naturalist, born A. D. 1614, translated Neri into our language in 1654, with many notes of his own about him; his observations have added nothing of value to the chemistry of inks.

Johann Kunckel, a noted German chemist and writer in 1657, republished in the German language Neri's work with Merret's notes, and his own observations on both. He also inserted many other processes as the result of considerable research and seems to have been thoroughly conversant with the chemistry of inks, advocating especially the value and employment of a tanno-gallate of iron ink for record purposes.

Salmon, A. D. 1665, in his Polygraphics, proceeds to give instructions relative to inks which notwithstanding their merit are confounded with so many absurdities as to lessen their value for those who were unable to separate truth from falsehood; but he nevertheless dwells on the virtues of the "gall" inks.

Jacques Lemort, a Dutch chemist of some note, issued a treatise, A. D. 1669, on "Ink Formulas and Colors," seemingly selected from the books of those who had preceded him. He expresses the opinion that the "gall" inks if properly compounded would give beneficial results.

Formulas for making inks are found tucked away in some of the very old literature treating of "curious" things. One of them which appeared in 1669 directs: "to strain out the best quality of iron employ old and rusty nails;" another one says, that the ink when made is to remain in an open vessel "for thirty days and thirty nights, before putting it in a parchment bag."

An English compendium of ink formulas, published in 1693, calls attention to many formulas for black inks as well as gold, silver, and the colored ones; no comment, however, is made in respect to any particular one being better than another as to permanency, and these conditions would seem to have continued for nearly a century later, though the art of handwriting was making giant strides.

It is a remarkable fact that notwithstanding the numerous devotees to that art which included many of the gentler sex, reproductions of whose skill in "Indian" ink are to be found engraved in magnificent publications, both in book and other forms, there is no mention in them or in any others included within this period about the necessity of using any other DURABLE ink for record or commercial purposes.

As indicative in some degree of the progress of the art of handwriting and handwriting materials, commencing A. D. 1525 and ending A. D. 1814, I present herewith a compilation of the names of over one hundred of the best known calligraphers and authors of the world, and not to be found as a whole in any public or private library. It is arranged in chronological order.

1525.

The first English essay on the subject of "Curious Calligraphy" was by a woman who from all accounts possessed most remarkable facility in the use of the pen as well as a knowledge of languages. Her name was Elizabeth Lucar; as she was born in London in 1510 and died 1537, her work must have been accomplished when only fifteen years of age.

1540.

Roger Ascham, best known as the tutor of
Queen Elizabeth.

1570.

Peter Bales, author of many works, "The Writing Schoolmaster," which he published in three parts, being the best known. He was also a microscopic writer. His rooms were at the sign of "The Hand and Golden Pen," London.

1571.

John de Beauchesne, teacher of the Princess Elizabeth, daughter of King James I. Author of many copy books.

1588.

John Mellis, "Merchants Accounts," etc.

1600.

Elizabeth Jane Weston, of London and Prague, wrote many poems in old Latin.

1600.

Hester Inglis, "The Psalms of David."

1601.

John Davies, "The Writing Schoolmaster, or
Anatomy of Fair Writing."

1616.

Richard Gething, "The Hand and Pen; 1645, "Chirographia" and many others.

1618.

Martin Billingsley, "The Writing Schoolmaster, or the Anatomie of Fair Writing." This author was writing master to King Charles I.

1622.

David Brown, who was scribe to King James I.
"Calligraphia."

1622.

William Comley, "Copy-Book of all the most usual English Hands," etc.

1646.

Josiah Ricrafte, "The Peculiar Character of the Oriental Languages."

1650.

Louis Hughes, "Plain and Easy Directions to
Fair Writing."

1650.

John Johnson, "The Usual Practices of Fair and Speedy Writing."

1651.

John Clithers, "The Pens Paradise," dedicated to Prince Charles.

1652.

James Seamer, "A Compendium of All the
Usual Hands Written in England."

1657.

Edward Cocker, penman and engraver, famous in his time for the number and variety of his productions. Author of "The Pen's Triumph," "The Artist's Glory," "England's Penman," and many more.

1659.

James Hodder, "The Penman's Recreation," etc.

1660.

John Fisher, "The Pen's Treasury."

1663.

Richard Daniel, "A Compendium of many hands of Various Countries."

1669.

Peter Story or Stent, "Fair Writing of Several Hands in Use."

1678.

William Raven, "An Exact Copy of the Court Hand."

1680.

Peter Ivers, famous for his engrossing and drawings.

1680.

Thomas Watson, "Copy-Book of Alphabets."

1681.

John Pardie, "An Essay on the German Text and Old Print Alphabets."

1681.

Thomas Weston, "Ancilla Calligraphiae."

1681.

Peter Gery, "Copy book of all the Hands in use, Performed according to the Natural Freeness of the Pen."

1681.

William Elder, "Copy-book of the most useful and necessary Hands now used in England."

1683.

John Ayers, "Tutor to Penmanship," and many others.

1684.

Caleb Williams, "Nuncius Oris," written and engraved by himself.

1693.

Charles Snell, "The Penman's Treasury Opened;" 1712, "Art of Writing in Theory and Practice;" 1714, "Standard Rules," etc.

1695.

Richard Alleine, writing master.

1695.

Eleazer Wigin, "The Hand and Pen."

1695.

John Sedden, "The Penman's Paradise."

1696.

John Eade, writing master.

1699.

Joseph Alleine, published several books about writing and accounts.

1699.

Robert More, "The Writing Masters Assistant." 1725. "The General Penman."

1700.

John Beckham, father of the celebrated George Beckham, wrote and engraved several pieces for "The Universal Penman."

1700.

Edward Smith, "The Mysteries of the Pen in fifteen Hands, Unfolded," etc.

1700.

Henry Legg, "Writing and Arithmetic."

1702.

William Banson, "The Merchants Penman."

1703.

John Dundas, microscopic writer.

1705.

George Shelley, "The Penmans Magazine."
In 1730 he wrote several pages for "Bickman's
Universal Penman."

1708.

John Clark, "The Penmans Diversion."

1709.

James Heacock, writing master.

1709.

George Shelley, "Natural writing in all hands."

1711.

George Bickham, one of the most famous of writers of his time, born 1684, died 1758, author of "The Universal Penman." He published many works. 1711, "The British Penman;" 1731, "Penmanship in its utmost Beauty and Extent" and "The Universal Penman" are the best known.

1709.

John Rayner, "Paul's Scholars Copy-Book."

1711.

Humphrey Johnson, "Youth's Recreation: a
Copy-Book of Writing done by Command of

Hand."

1712.

William Webster, writing and mathematics. 1730, wrote several pages for "The Universal Penman."

1713.

Thomas Ollyffe, "The Hand and Pen." 1714,
"The Practical Penman."

1717.

William Brooks, "Delightful Recreation for the Industrious." Contributor to "The Universal Penman."

1717.

Abraham Nicholas, "Various Examples of Penmanship." 1722, "The Compleat Writing Master." Wrote also for "The Universal Penman."

1719.

Ralph Snow, "Youths Introduction to Handwriting."

1720.

William Richards, "The Complete Penman."

1723.

John Jarman, "A System of Court Hands."

1724.

Henry Lune, "Round Hand Complete."

1725.

John Shortland, writing master and contributor to "The Universal Penman."

1725.

Edward Dawson, writing master and contributor to "The Universal Penman."

1726.

Moses Gratwick, writing master and contributor to "The Universal Penman."

1727.

John Langton, "The Italien Hand."

1728.

John Day, writing master and contributor to
"The Universal Penman."

1729.

Gabriel Brooks, writing master and contributor to, "The Universal Penman."

1730.

William Keppax, writing master and contributor to "The Universal Penman."

1730.

John Bland, "Essay in Writing." Also contributor to "The Universal Penman."

1730.

Solomon Cook, "The Modish Round Hand."

1730.

William Leckey, "A Discourse on the Use of the Pen." Contributor to "The Universal Penman."

1730.

Peter Norman, writing master and contributor to "The Universal Penman."

1730.

Wellington Clark, writing master and contributor to "The Universal Penman."

1730.

Zachary Chambers, "Vive la Plume." Contributor to "The Universal Penman."

1733.

Bright Whilton, writing master and contributor to "The Universal Penman."

1734.

Timothy Treadway, writing master and contributor to "The Universal Penman."

1738.

George J. Bickham, writing master; also wrote for "Bickham's Universal Penman."

1739.

Emanuel Austin, writing master; he wrote 22 pages in "The Universal Penman."

1739.

Samuel Vaux, writing master and contributor to "The Universal Penman."

1740.

Jeremiah Andrews, writing master and tutor to King George III.

1740.

Nathaniel Dove, "The Progress of Time," and contributor to "The Universal Penman."

1741.

John Blande, "Essay in Writing; 1730, contributor to "The Universal Penman."

1741.

Richard Morris, writing master and contributor to "The Universal Penman."

1747.

Mary Johns, microscopic writer and author.

1749.

Charles Woodham, "A Specimen of Writing, in the most Useful Hands now Practised in England."

1750.

John Oldfield, "Honesty." He wrote one piece in "The Universal Penman."

1750.

Joseph Champion, "The Parallel or Comparative Penmanship." 1762, "The Living Hands."

1751.

Edward Lloyd, "Young Merchants Assistant."

1758.

Richard Clark, "Practical and Ornamental Penmanship."

1760.

Benjamin Webb, writer of copy books, etc.

1762.

William Chinnery, "The Compendious Emblematist."

1763.

William Massey, "The Origin and Progress of Letters," containing valuable information about the art.

1769.

John Gardner, "Introduction to the Counting House."

1780.

Edward Powell, writing master and designer.

1784.

E. Butterworth, "The Universal Penman" in two parts, published in Edinburgh.

1795.

William Milns, "The Penman's Repository."

1799.

William G. Wheatcroft, "The Modern Penman."

1814.

John Carstairs, "Tachygraphy, or the Flying Pen." 2. "Writing made easy, etc."

Illustrated works on the subject of penmanship of contemporaneous times and not of English origin are but few. The best known are:

1543.

Luduvico Vicentino, "A Copy book" published in Rome, seems to have been the first.

1570.

Il perfetto Scrittore (The Perfect Writer) by Francesco Cresci, published in Rome.

1605.

Spieghel der Schrijkfkonste (or Mirror of Penmanship) written by Van den Velde, published in Amsterdam.

1612.

"Writing and Ink Recipes," by Peter Caniparius, Venice and London.

1700.

Der Getreue Schreibemeister (or True Writing Master), by Johann Friedr Vicum, published in Dresden.

From 1602 to 1709 many "Indian" ink specimens were extant and are still of the different schools of penmanship. The productions of Phrysius, Materot and Barbedor illustrating the French style, Vignon, Sellery and others, for the Italian hand, and Overbique and

Smythers for the German text, and Ambrosius Perlengh and Hugo, with a few more, complete the list.

CHAPTER XII.

STUDY OF INK.

LACK OF INTEREST AS TO THE COMPOSITION OF INK DURING PART OF THE EIGHTEENTH CENTURY—THE CONDITIONS WHICH THEN PREVAILED NEARLY THE SAME AS THE PRESENT TIME—CHEMISTRY OF INK NOT UNDERSTOOD— THIS LACK OF INFORMATION NOT CONFINED TO ANY PARTICULAR COUNTRY—LEWIS, IN 1765, BEGINS A SCIENTIFIC INVESTIGATION ON THE SUBJECT OF INKS —THE RESULTS AND HIS CONCLUSIONS PUBLISHED IN 1797—THE ROYAL SOCIETY OF ENGLAND IN 1787 RECEIVES COMPLAINTS ABOUT THE INFERIORITY OF INKS —ITS SECRETARY READS A PAPER THE SAME YEAR—THE PAPER CITED IN FULL—DR. BOSTOCK IN 1830 COMMUNICATES TO THE SOCIETY OF ARTS WHAT HE ESTIMATES TO BE THE CAUSES OF IMPERFECTIONS IN INK—ACTION OF THE FRENCH ACADEMY OF SCIENCES— COMPLICATIONS SURROUNDING THE MANUFACTURE OF INK ONLY THIRTY-FIVE YEARS AGO.

THE increasing demands for ink, and the lack of interest as to its composition during the eighteenth century, if viewed in the same lights which prevail in our own times, permitted the general manufacture of cheap grades of ink which possessed no very lasting qualities. The chemistry of Inks was not fully understood, indeed we find Professer Turner of the College of Edinburgh declaring in 1827:

"Gallic acid was discovered by Scheele in 1786, and exists ready formed in the bark of many trees, and in gall-nuts. It is always associated with tannin, a substance to which it is allied in a manner hitherto unexplained. It is distinguished from tannin by causing no precipitate in a solution of gelatine. With a salt of iron it forms a dark blue coloured compound, which is the basis of ink. The finest colour is procured when the peroxide and protoxide of iron are mixed together. This character distinguishes gallic acid from every other substance excepting tannin."

The general lack of information or knowledge respecting ink chemistry or its time-phenomena was not confined to any particular country, and it does not appear that any general or specific attention was scientifically directed to it until 1765, when William Lewis, F. R. S., an English chemist, publicly announced that he proposed to investigate the subject. His experimentations covered a period of many years and their results and his theories as to the phenomena of inks were published in 1797. The most valuable of his conclusions were that an excess of iron salt in the ink is detrimental to color permanence (such ink becoming brown on exposure) and also that acetic acid in the menstruum provides an ink of greater body and blackness than sulphuric acid does (a circumstance due to the smaller resistance of acetic acid to the formation of iron gallo-tannate). Many of his other observations were later shown to have been erroneous. Dr. Lewis was the first to advocate log- wood as a tinctorial agent in connection with iron and gall compositions.

Ribaucourt, a French ink maker, in 1798 determined that an excess of galls is quite as injurious to the permanence of ink as an excess of iron.

Pending the completion of the researches of Lewis, the Royal Society of England, affected by complaints from all quarters relative to the inferiority of inks as compared with those of earlier times, brought the subject to the attention of many of its members for discussion and advice. Its secretary, Charles Blagden, M. D., read a paper before the society, June 28, 1787, which was published in the "Philosophical Transactions" and widely circulated. It is so interesting that copious extracts are given:

"In a conversation some time ago with my friend Thomas Astle, Esq., F. R. S. and A. S., relative to the legibility of ancient MSS. a question arose, whether the inks in use eight or ten centuries ago, which are often found to have preserved their colour remarkably well, were made of different materials from those employed in later times, of which many are already become so pale as scarcely to be read. With a view to the decision of this question, Mr. Astle obligingly furnished me with several MSS., on parchment and vellum, from the ninth to the fifteenth centuries inclusively, some of which were still black, and others of different shades of colour, from a deep yellowish brown to a very pale yellow, in some parts so faint as to be scarcely visible. On all of these I made experiments with the chemical re-agents which appeared to me best adapted to the purpose, namely, alkalis both simple and phlogisticated, the mineral acids, and infusions of galls.

"It would be tedious and superfluous to enter into a detail of the particular experiments, as all of them, one instance only excepted, agreed in the general result, to shew that the ink employed anciently, as far as the above-mentioned MSS. extended, was of the same nature as the present; for the letters turned of a reddish or yellow brown with alkalis, became pale, and were at length obliterated, with the dilute mineral acids, and the drop of acid liquor which had extracted a letter, changed to a deep blue or green on the addition of a drop of phlogisticated alkali; moreover, the letters acquired a deeper tinge with the infusion of galls, in some cases more, in others less. Hence it is evident, that one of the ingredients was iron, which there is no reason to doubt was joined with the vitriolic acid; and the colour of the more perfect MSS. which in some was deep black, and in others purplish black, together with the restitution of that colour, in those which had lost it, by the infusion of galls, sufficiently proved that another of the ingredients was a stringent matter, which from history appears to be that of galls. No trace of a black pigment of any sort was discovered, the drop of acid which had completely extracted a letter, appearing of an uniform pale ferrugineous color, without an atom of black powder, or other extraneous matter, floating in it.

"As to the durability of the more ancient inks, it seemed, from what occurred to me in these experiments, to depend very much on a better preparation of the material upon which the writing was made, namely, the parchment or vellum; the blackest letters being those which had sunk into it deepest. Some degree of effervescence was commonly to be perceived when the acids came into contact with the surface of these old vellums. I was led, however, to suspect, that the more modern; for in general the tinge of colour, produced by the phlogisticated alkali in the acid laid upon them, seemed less deep; which, however, might depend in part upon the length of time they have been kept: and perhaps more gum was used in them, or possible they were washed over with some kind of varnish, though not such as gave gloss.

"One of the specimens sent me by Mr. Astle, of the fifteenth century, and the letters were those of an engrossing hand, angular, without any FINE strokes, broad and very black. On this none of the above-mentioned re-agents produced any considerable effect; most of them seemed to make the letters blacker, probably by cleaning the surface; and the acids, after having been rubbed strongly on the letters, did not strike any deeper tinge with the phlogisticated alkali. Nothing had a sensible effect toward obliterating these letters but what took off part of the surface of the vellum, when small rolls, as of a dirty matter, were to be perceived. It is therefore unquestionable, that no iron was used in this ink; and from its resistance to the chemical solvents, as well as a certain clotted appearance in the letters when examined closely, and in some places a slight degree of gloss, I have little doubt but they were formed with a composition of a black, sooty or carbonaceous powder and oil, probably something like our present printer's ink, and am not without suspicion that they were actually printed (a subsequent examination of a larger portion of this supposed MSS. has shown that it is really a part of a very ancient printed book).

"Whilst I was considering of the experiments to be made, in order to ascertain the composition of ancient inks, it occurred to me that perhaps one of the best methods of restoring legibility to decayed writing might be to join phlogisticated alkali with the remaining calx of iron, because, as the quantity of precipitate formed by these two substances very much exceeds that of the iron alone, the bulk of the colouring matter would thereby be greatly augmented. M. Bergman was of opinion that the blue precipitate contains only between a fifth and a sixth part of its weight of iron, and though subsequent experiments tend to show that, in some cases at least, the proportion of iron is much greater, yet upon the whole it is certainly true, that if the iron left by the stroke of a pen were joined to the colouring matter of phlogisticated alkali, the quantity of Prussian blue thence resulting would be much greater than the quantity of black matter originally contained in the ink deposited by the pen, though perhaps the body of colour might not be equally augmented. To bring the idea to the test, I made a few experiments as follows:

"The phlogisticated alkali was rubbed upon the bare writing in different quantities, but in general with little effect. In a few instances, however, it gave a bluish tinge to the letters, and increased their intensity, probably where something of an acid nature had contributed to the diminution of their colour.

"Reflecting that when phlogisticated alkali forms its blue precipitate with iron the metal is first usually dissolved in an acid, I was next induced to try the effect of adding a dilute mineral acid to writing besides the alkali. This answered fully to my expectations, the letters changing very speedily to a deep blue colour, of great beauty and intensity.

"It seems of little consequence as to the strength of colour obtained, whether the writing be first wetted with the acid, and then the phlogisticated alkali be touched upon it, or whether the process be inverted, beginning with the alkali; but on another account I think the latter way preferable. For the principal inconvenience which occurs in the proposed method of restoring MSS. is, that the colour frequently spreads, and so much blots the parchment as to detract greatly from the legibility; now this appears to happen in a less degree when the alkali is put on first, and the dilute acid is added upon it.

"The method I have hitherto found to answer best has been to spread the alkali thin with a feather or a bit of stick cut to a blunt point, though the alkali has occasioned no sensible change of colour, yet the moment that the acid comes upon it, every trace of a letter turns at once to a fine blue, which soon acquires its full intensity, and is beyond comparison stronger than the colour of the original trace had been. If now the corner of a bit of blotting paper be carefully and dexterously applied near the letters, in order to suck up the superfluous liquor, the staining of the parchment may be in a great measure avoided: for it is this superfluous liquor which absorbing part of the colouring matter from the letters becomes a dye to whatever it touches. Care must be taken not to bring the blotting paper in contact with the letters, because the colouring matter is soft whilst wet, and may easily be rubbed off. The acid I have chiefly employed has been the marine; but both the vitriolic and nitrous succeed very well. They should undoubtedly be so far diluted as not to be in danger of corroding the parchment, after which the degree of strength does not seem to be a matter of much nicety.

"The method now commonly practiced to restore old writings, is by wetting them with an infusion of galls in white wine."

(See a complicated process for the preparation of such a liquor in Caneparius De Atramentis, A. D. 1660, p. 277)

"This certainly has a great effect; but is subject, in some degree, to the same inconvenience as the phlogisticated alkali, of staining the substance on which the writing was made. Perhaps if, instead of galls themselves, the peculiar acid of or other matter which strikes the black with iron were separated from the simple astringent matter, for which purpose two different processes are given by Piesenbring and by Scheele, this inconvenience might be avoided. It is not improbable, likewise, that a phlogisticated alkali might be prepared better suited to this object than the common; as by rendering it as free as possible from iron, diluting it to a certain degree, or substituting the volatile alkali for the fixed. Experiment would most likely point out many other means of improving the process described above; but in its present state I hope it may be of some use, as it not only brings out a prodigious body of colour upon letters which were before so pale as to be almost invisible, but has the further advantages over the infusions of galls, that it produces its effect immediately, and can be confined to these letters only for which such assistance is wanted."

The Society of Arts in 1830, received a communication from Dr. Bostock, in the course of which he stated that the "tannin, mucilage and extractive matter are without doubt the principal causes of the difficulty which is encountered in the formation of a perfect and durable ink and for a good ink the essential ingredients are gallic acid and a sesqui salt of iron." Owing to his working with galls he was unable to make decisive experiments, but he concludes, and that rightly, that in proportion as ink consists merely of gallate of iron, it is less liable to decomposition and any kind of metamorphosis.

In 1831 the Academy of Sciences in France took up the matter and designated a committee composed of chemists with instructions to study the subject of a permanent ink. After long research it reported that it was unable to recommend any better ink than the tanno-gallate of iron one then in use, but "it should be properly compounded."

Peddington investigated, 1841-48, the ancient MSS. collected by the Asiatic Society of Bengal, Calcutta, and published the results in "Examination of Some Decayed Oriental Works in the Library of the Asiatic Society," which are of much interest as relating to "mineral" inks, the "gall" inks being unknown in Asia after the twelfth century.

Up to thirty-five years ago, the manufacture of "gall" inks necessitated a complicated series of processes and long periods of time to enable the ink to settle properly, etc. It was Professor Penny of the Anderson University who suggested the way to avoid one of the processes pertaining to ink-making by utilizing the known fact, that tannin is more soluble in cold than in warm or hot water. It was adopted all over the world and revolutionized the manufacture of ink, by doing away with boiling processes and hot macerations of ingredients. With hardly in exception the best tanno-gallate of iron ("gall") inks are now "cold" made.

CHAPTER XIII.

STUDY OF INK.

INVESTIGATIONS BY STARK OF INK QUALITIES COVERING A PERIOD OF TWENTY-THREE YEARS—ABSTRACT FROM HIS REPORT OF 1855—DR. CHILTON EXPERIMENTS IN NEW YORK CITY 1856—ACTION OF THE PRUSSIAN GOVERNMENT IN 1859 AND EMPLOYMENT OF AN OFFICIAL INK—WATTENBACH'S GERMAN TREATISE ON THE ARCHIVES OF THE MIDDLE AGES—WILLIAM INGLIS CLARK ATTEMPTS TO PLACE THE MANUFACTURE OF INK ON A SCIENTIFIC BASIS—SUBMITS HIS VALUABLE RESEARCHES

AND DEDUCTIONS TO THE ENINBURGH UNIVERSITY IN 1879—SCHLUTTIG AND NEUMANN IN 1890 ESTABLISH A STANDARD FORMULA FOR IRON AND GALL INK—NAMES OF SOME INK INVESTIGATORS OF THE NINETEENTH CENTURY.

DR. JAMES STARK, a famous chemist, submitted the results of twenty-three years of investigations of writing inks in a paper read by him in 1855 before the Society of Arts, in Edinburg, Scotland. The following is the abstract as printed by the London Artisan at the time:

"The author stated that in 1842 he commenced a series of experiments on writing inks, and up to this date (1855), had manufactured 229 different inks, and had tested the durability of writings made with these on all kinds of paper. As the result of his experiments be showed that the browning and fading of inks resulted from many causes, but in ordinary inks chiefly from the iron becoming peroxygenated and separating as a heavy precipitate. Many inks, therefore, when fresh made, yielded durable writings; but when the ink became old, the tanno-gallate of iron separated, and the durability of the ink was destroyed. From a numerous set of experiments the author showed that no salt of iron and no precipitate of iron equalled the common sulphate of iron—that is, the commercial copperas—for the purpose of ink- making; and that even the addition of any persalt, such as the nitrate or chloride of iron, though it improved the present color of the ink, deteriorated its durability. The author failed to procure a persistent black ink from manganese, or other metal or metallic salt. The author exhibited a series of eighteen inks which had either been made with metallic iron or with which metallic iron had been immersed, and directed attention to the fact that though the depth and body of color seemed to be deepened, yet in every case the durability of writings made with such inks was so impaired that they became brown and faded in a few months. The most permanent ordinary inks were shown to be composed of the best blue gall nuts with copperas and gum, and the proportions found on experiment to yield the most persistent black were six parts of best blue galls to four parts of copperas. Writings made with such an ink stood exposure to sun and air for twelve months without exhibiting any change of color, while those made with inks of every other proportion or composition had more or less of their color discharged when similarly tested. This ink, therefore, if kept from moulding and from depositing its tanno-gallate of iron, would afford writings perfectly durable. It was shown that no gall and logwood ink was equal to the pure gall ink in so far as durability in the writings was concerned. All such inks were exhibited which, though durable before the addition of logwood, faded rapidly after logwood was added to them. Sugar was shown to have an especially hurtful action on the durability of inks containing logwood—indeed, on all inks. Many other plain inks were exhibited, and their properties described —as gallo-sumach ink, myrabolams ink, Runge's ink, —inks in which the tanno-gallate of iron was kept in solution by nitric, muriatic, sulphuric, and other acids, or by oxalate of potash, chloride of lime, etc. The myrabolams was recommended as an ink of some promise for durability, and as the cheapest ink it was possible to manufacture. All ordinary inks, however, were shown to have certain drawbacks, and the author endeavored to ascertain by experiment whether other dark substances could be added to inks to impart greater durability to writings made with them, and at the same time prevent those chemical changes which were the cause of ordinary inks fading. After experimenting with various substances, and among others with Prussian blue and indigo dissolved in various ways, he found the sulphate of indigo to fulfil all the required conditions and, when added in the proper proportion to a tanno-gallate of iron ink, it yielded an ink which is agreeable to write with, which flows freely from the pen and does not clog it; which never moulds, which, when it dries on the paper, becomes of an intense pure black, and which does not fade or change its color however long kept. The author pointed out the proper proportions for securing those properties, and showed that the smallest quantity of the sulphate of indigo which could be used for this purpose was eight ounces for every gallon of ink. The author stated that the ink he preferred for his own use was composed of twelve ounces of gall, eight ounces of sulphate of indigo, eight ounces of copperas, a few cloves, and four or six ounces of gum arabic, for a gallon of ink. It was shown that immersing iron wire or filings in these inks destroyed ordinary inks. He therefore recommended that all legal deeds or documents should be written with quill pens, as the contact of steel invariably destroys more or less the durability of every ink. The author concluded his paper with a few remarks on copying inks and indelible inks, showing that a good copying ink has yet to be sought for, and that indelible inks, which will resist the pencilings and washings of the chemist and the forger, need never be looked for."

Professor Leonhardi, of Dresden, who had given much attention to the subject of inks, introduced in 1855 what he termed a NEW ink, and named it "alizarine ink," alizarin being a product obtained from the madder root, which he employed for "added" color in a tanno-gallate of iron solution. It possessed some merit due to its fluidity, and for a time was quite popular, but gradually gave place to the so-called chemical writing fluids; it is now obsolete.

Champour and Malepeyre, Paris, 1856, issued a joint manual, "Fabrication des Encres," devoted almost exclusively to the manufacture of inks and compiles many old "gall" and other ink formulas.

In 1856 Dr. Chilton of New York City published the results of ink experiments which he had made. The accompanying extracts are taken from the local press of the month of April of that year:

"Some ingenious experiments to test the durability of writing inks have recently been made by Dr. Chilton, of New York City. He exposed a manuscript written with four different inks of the principal makers, of this and other countries, to the constant action of the weather upon the roof of his laboratory. After an exposure of over five months, the paper shows the different kind of writing in various shades of color. The English sample, Blackwood's, well known and popular from the neat and convenient way that it is prepared for this market, was quite indistinct.

"The American samples, David's, Harrison's and Maynard's are better. The first appears to retain its original shade very neatly; the two last are paler. This test shows conclusively the durability of ink; and while, for many purposes, school and the like, an ink that will stand undefaced a year or so, is all that is necessary, yet there is hardly a bottle of ink sold, some of which may not be used in the signature or execution of papers that may be important to be legible fifty or one hundred years hence.

"For state and county offices, probate records, etc., it is of vital importance that the records should be legible centuries hence. We believe that some of the early manuscripts of New England are brighter than some town and church records of this century.

"In Europe at the present time, great care is taken by the different governments in the preparation of permanent ink—some of them even compounding their own, according to the most approved and expensive formulas.

"Manuscripts of the eleventh and twelfth centuries now in the state paper office of Great Britain, are apparently as bright as when first written; while those of the last two hundred years are more or less illegible, and some of them entirely obliterated."

While the information sought to be conveyed in the last statement may be in some respects correct, it must be remembered that most of the MSS. extant dating before the thirteenth century were written in "Indian" ink, while the great majority of those of the last two hundred years were not; and this fact alone would account to some extent for the differences mentioned.

The German (Prussian) government in 1859, as the result of an investigation, employed what they termed "Official Ink of the First Class," i. e., a straight tanno- gallate of iron ink without added color; and if permanence were required as against removal by chemicals, it was accomplished by writing on paper saturated with chromates and ultramarine.

In 1871 Professor Wattenbach of Germany published a treatise entitled "Archives during the Middle Ages," which has some valuable references to the color phenomena of inks.

William Inglis Clark in 1879 submitted to the Edinburgh University a thesis entitled "An Attempt to Place the Manufacture of Ink on a Scientific Basis," and which very justly received the commendation of the University authorities. His researches and rational deductions are of the greatest possible value judged from a scientific standpoint. The introduction of blue-black ink is a phase of the development towards modern methods which he discusses at much length.

The object of adding a dye in moderation, he asserts, is to give temporary color to the ink and where indigo-paste is used, it has been assumed that it kept the iron gallo-tannate in solution, whereas any virtue of this kind which indigo-paste possesses is more likely due to the sulphuric acid which it contains than to the indigo itself. The essential part of the paste required is the sulpho-indigodate of sodium, now commonly called indigo-carmine. He further remarks that the stability of an ink precipitate depends upon the amount of iron which it contains and which on no account should be less than eight per cent; he adds rightly, if gallic acid be preferably used in substitution for tannin, "no precipitate is obtained under precisely similar conditions." This point followed up explains in a measure why a gall infusion prepared with hot water is not suitable for a blue-black, while a cold water infusion is. In the latter case a comparatively small percentage of tannin is extracted from the galls, while much is extracted with hot water and the consequence is, on adding the indigo blue the color is not brought out as it should be. Substantially the same thing occurs with ink made with the respective acids, although the blue color remains for a time unimpaired in the tannin ink, apparently due to the fact that ferrous-tannate reduces indigo blue to indigo white, a change which the low reducing power of ferrous- gallate does little to effect. The vegetable matter in common inks facilitates the destruction, or rather alteration and precipitation of the indigo, for the dye appears in the iron precipitate and may be extracted from it with boiling water.

Dr. Clark's investigations seek to demonstrate the superiority of tannin and gallic acid over infusions of the natural galls, and he undertakes to determine the correct ratio of tannin and sulphate of iron to be used as ink. His experiments in this line show that:

1. The amount of precipitate increases as the proportion of iron to tannin is increased.

2. The composition of the precipitate is so valuable as to preclude the possibility of its being a definite body. Increase of iron in the solution has not at first any effect on the composition of the precipitate, but afterwards iron is found in it in greater but not proportional amount.

3. At one point the proportions of iron in the precipitate and in solution are the same, and this is at between 6 and 10 parts of iron to 100 parts of tannin.

4. The proportion of iron in the precipitate varies greatly with the length of time the ink has been exposed. At first the precipitate contains 10 per cent of iron, but by and by a new one having only 7.5 per cent is formed, and in from forty to seventy days we find one of 5.7 per cent. Simultaneously iron increases in the ink (proportionate to the tannin).

5. The results show, and practice confirms, that 16 parts of iron (80 ferrous sulphate) and 100 parts of tannin are best for ink manufacture.

The research now travelled in a direction which accumulating experience showed to be obligatory. Blue-black tannin ink lost color, and the reducing nature of the tannin tended to the formation of a highly objectionable precipitate in the ink, which made writing anything but a pleasure. These two faults were doubtless linked together in some way and seemed not to exist when gallic acid was used, for ink so made was found to precipitate only after a long exposure, it required no free acid to keep the precipitate in solution, and retained the indigo blue color for a long time; alkalis did not decompose the ink, and provided blacker and more permanent writing. Determination of the correct proportions of gallic acid and ferrous-sulphate was the subject of prolonged experiments conducted on similar lines to those already detailed. The conclusions as to precipitation were also similar. Thirty parts of iron (150 of ferrous- sulphate) and 100 parts of gallic acid were found to be the most suitable proportions for ink-making. It is advisable, however, not to discard tannin altogether, owing to the slow blackening of the gallic acid ink, and a little tannin gives initial blackening and body, while it is absolutely necessary for copying ink. Initial blackness can also be ensured by oxidizing 21 per cent of the ferrous-sulphate without adding the extra acid necessary to the formation of a ferric salt.

The concluding portion of his research is devoted to the influence of sugar upon the permanence of ink, and the results of the experiments are summed up in the following sentences: "It would be injurious to add 3 per cent of sugar to a tan in ink, while from 4 to 10 per cent would be quite allowable. Most copying inks contain about 3.5 per cent of sugar— not far from the critical amount. With gallic acid more than 3 per cent of sugar hardly varies the precipitate, but the importance of this point is somewhat diminished by the fact that the presence of sugar is by no means necessary in a writing ink. Dextrin is a much superior substance to use. Curiously this body rapidly precipitates a tannin ink; hence it is useless for copying ink, but for the gallic ink it is an excellent thickener."

Chen-Ki-Souen, "Lencre de China," by Maurice Jametel, appeared in Paris in 1882, but as the title indicates, it is the old "Indian" or Chinese ink that is discussed.

Schluttig and Neumann in 1890 issued their Edition Dresden on the subject of "Iron and Gall inks." In this valuable work is to be found the formula which has been generally adopted as the standard where one is used for tanno-gallate of iron ink.

The investigations of other scientific men like Lepowitz, Booth, Desormeaux, Chevreuse, Irvine, Traille, Bottger, Riffault, Precht, Nicholes, Runge, Gobert, Penny, Arnold, Thomson (Lord Kelvin), Davids, Kindt, Ure, Wislar and many more who have dealt with the chemistry of inks, present to us some testimony during a considerable portion of the nineteenth century of the efforts made to secure a good ink.

CHAPTER XIV.

CLASSIFICATIONS OF INK.

INK USED BY US HAS NOTHING IN COMMON WITH THAT OF THE ANCIENTS—MANUFACTURERS OF THE PRESENT TIME HAVE LARGELY UTILIZED FORMULAS EMPLOYED IN PAST CENTURIES—THE COMMON ACCEPTATION OF THE TERM INK—SEVEN DIFFERENT CLASSES OF INKS AND THEIR COMPOSITION BRIEFLY TOLD—FAILURE OF EFFORTS TO SECURE A REAL SAFETY INK.

THE inks used by us have nothing in common with those of the ancients except the color and gum, and mighty little of that.

Those of the "gall" class employed in the fourteenth, fifteenth, sixteenth, seventeenth and eighteenth centuries, some formulas of which are utilized by the manufacturers of ink in our own time, consisted generally in combination; infusions of nut-galls, sulphate of copper or iron, or both, and fish-glue or gum, slightly acidulated. The frequent introduction of the so-called "added" color into these inks, time has shown to have been a grave mistake.

The common acceptation of the term "ink" may be said to characterize an immense number of fluid compounds, the function of which in connection with a marking instrument is to delineate conventional signs, characters and letters as put together and commonly called writing, on paper or like substances.

To classify them would be impossible; but black writing ink, chemical writing fluid, colored writing ink, copying ink, India ink, secret or sympathetic ink, and indelible ink make seven classes; the others may be denominated under the head of miscellaneous inks, and of them all, there is no single ink answering every requirement and few answer at all times the same requirements. Ink may be either a clear solution of any coloring matter or of coloring matter held in suspension. It is a remarkable fact that although most inks are chemical compositions and many times made after the same formula, identical results cannot always be calculated or obtained. This is more particularly to be noted in the case of black writing inks otherwise known as the tanno-gallate of iron inks [gallic and gallotanic acid obtained from nut-galls, sulphate of iron, (green copperas) and some gummy vehicle].

The variations would appear to be largely due to the difference in quality of the gall-nuts, treatment, and temperature of the atmosphere; perhaps, however, not so much to-day as it was ten or twenty years ago, when to make ink of this character boiling processes were employed. Most of them as already stated are now "cold" made.

Inks of this class consist of a finely divided insoluble precipitate suspended in water by the use of gum and possessing a slight acidity.

The requisites of a good black writing ink or black writing fluid require it to flow readily from the pen, to indicate in a short time a black color and to penetrate the paper to an appreciable degree, and more important than all the rest, to be of great durability. When kept in a closed vessel no sediment of any account should be precipitated, although such will be the case in open ink-wells, and this the quicker the more the air is permitted to get to it. If it is to be used for record or documentary purposes it must not be altogether obliterated if brought into contact with water or alcohol, and should depend for permanency on its chemical and not on its pigmentary qualities.

The second class, called for distinction "chemical writing fluids," possesses the same essential ingredients to be found in class one, but much less in quantity and with some "added" colored substance which I shall term "loading," for its real purpose is to cheapen the cost of production and not altogether as some manufacturers state "simply to give them an agreeable color."

Previous to the discovery of the soluble anilines, logwood, indigo, madder, orchil and other dyeing materials were used for a period of some eighty years and vanadium for some twenty years (very costly at that time), for this purpose, but since 1874, and with frequent changes as the newer aniline compounds were invented, these by-products of coal-tar, as well as logwood, etc., have been and are to-day employed for "loading," or as the manufacturer expresses, it "added color." The chemical writing fluids as now prepared, yield when first written a blue or green color with a tendency to change to black afterwards. They are not as permanent as those of the first class.

Another black ink not durable, however, is "logwood;" its extract is combined with a little chromate of potassium and boiled together in water. It possesses its own "gum" and contains some tannin. In combination with alum and water, it forms a dark purple ink.

The colored writing inks, of which "red" is the more important, are in great number and with hardly an exception at the present time, manufactured by adding water and water-glass to a soluble aniline red color. Cochineal which was used for red ink formerly is now almost obsolete. Nigrosine, one of the best known of them, is much used as a cheap "black" ink, but as it is blue black and never becomes black, it really belongs to the family of "colored" writing inks. They possess an undeserved popularity for they flow freely from the pen which they do not corrode, nor do they thicken or spoil in the inkwell; they are however very "fugitive" in character and should not be employed for record, legal, monetary or other documentary purposes. The indigo and prussian blue inks are well known, the former under certain conditions a very permanent ink, the latter soon disintegrating.

Copying inks are of two kinds, one dependent on the addition of glycerine, sugar, glucose or like compounds to the black writing inks or chemical writing fluids heretofore mentioned, which are thereby kept in a moist offsetting condition; the other due to the solubility of the pigmentary color with water, such as the aniline inks which are given more body than those for ordinary purposes— and the logwoods in which the pigment is developed and given copying qualities by chemicals, and hence becomes responsive to the application of a sheet of paper dampened with water. Copying ink should never be used for "record" purposes as it is affected by changes of the temperature.

India ink, sometimes called China ink, or as formerly known by the ancients and in classical and later times "Indian ink," is now used more for drawing and engrossing than it is for commercial purposes. It belongs to the "carbon" class and in some form was the first one used in the very earliest times. In China it is applied with a brush or pith of some reed to the "rice" paper also there manufactured. It is easily washed away unless bichromate of ammonium or potassium in minute quantities be added to it, and then if the paper on which it appears be exposed for a short time to the action of the actinic rays of sunlight, this gummy compound will be rendered insoluble and cannot be removed with any fluid, chemical or otherwise. It possesses also great advantages in drawing, since it acts as a paint, and will give any degree of blackness according to the quantity of water mixed with it.

Secret or sympathetic inks are invisible until the writing is subjected to a subsequent operation, such as warming or exposing to sunlight. To further aid the object in view, the paper may be first steeped in a liquid and the writing only made visible by using another liquid which has some chemical affinity with the previous one. The number of this kind were but few but have multiplied as chemistry progressed. The ancients were acquainted with several modes. Ovid indiscreetly advises the Roman wives and maidens if they intend to make their correspondence unreadable to the wrong persons to write with new milk, which when dried may be rendered visible by rubbing ashes upon it or a hot iron. Pliny suggests milky juices of certain plants of which there are a considerable variety.

Indelible ink is not used for writing purposes on paper, but is found best adapted for marking linen and cancellation or endorsing purposes. It is chiefly composed of nitrate of silver preparations, to which heat must be applied after it has been dried; or a pigment is commingled with the same vehicles used in making common printing ink and in its use treated as such.

Diamonds, gold, silver, platinum and a host of other materials are manufactured into ink and are to be placed under the head of miscellaneous inks. They are in great number and of no interest in respect to ink writing except for engrossing or illuminating.

Still another ink once held in much esteem and now almost obsolete is the so-called "safety" ink.

Manufacturers, chemists and laymen in great number for many years wasted money, time and energy in diligent worship at a secret shrine which could not give the information they sought. A sum-mary of the meager and barren results they secured is of little value and unimportant. Hence, there is no REAL "safety" ink.

It is true that lampblack (carbon) as made into ink, resists any chemical or chemicals, but simple water applied on a soft sponge will soon remove such ink marks. The reason for this is obvious, the ink does not penetrate the paper.

"Safety" ink which will not respond to acids may be affected by alkalis, or if resisting them separately, will yield to them in combination.

CHAPTER XV.

OFFICIAL AND LEGAL INK.

FIRST COMPLETE OFFICIAL INVESTIGATION OF INK IN THIS COUNTRY—THE HONOR DUE TO ROBERT T. SWAN OF BOSTON—RESUME OF HIS REPORTS TO THE LEGISLATURE OF THE STATE OF MASSACHUSETTS—THE SWAN LAW ADOPTED IN 1894 BY THE STATE OF MASSACHUSETTS—UNITED STATES TREASURY DEPARTMENT ADOPTS AN OFFICIAL INK IN 1901—UNSUCCESSFUL ATTEMPT TO SECURE INK LEGISLATION IN THE STATE OF NEW YORK—COMMENTS OF THE PUBLIC PRESS OF THAT PERIOD—DIFFERENT WORKS WHICH MORE OR LESS DWELL ON THE SUBJECT OF INK FROM 1890 TO 1900—CITATIONS FROM ALLEN'S COMMERCIAL ORGANIC ANALYSIS—REFERENCE TO PAPER ABOUT INK READ BEFORE THE NEW YORK STATE BAR ASSOCIATION.

IT was not, however, until 1891 that the subject of the constitution of an enduring record ink received the consideration its importance deserved and in this the youngest of countries. To Robert T. Swan of Boston is all honor due for the very unique and comprehensive methods adopted in his investigations. Appointed "commissioner of public records" of the state of Massachusetts, he has set an example which may well be followed by other states, as has been done in a lesser degree by Connecticut and ten years later by the United

States Treasury Department, which in this respect is so ably represented in part by Dr. Charles A. Crampton of Washington, D. C.

Mr. Swan in his reports to the legislature of his state for the last twelve years, deals with the subject of the constitution of "permanent inks" so thoroughly, and with it affords information of so practical and useful a character, that the fullest references to them prove both instructive and interesting. In his report of 1891 he remarks:

"Upon commencing an examination of the records in various places, I was impressed with the great importance of the use of inks which should be permanent, and the necessity of an investigation which might prevent the further use of inks that for one reason or another were unfit for use upon records. I found that, as a rule, the inks upon the most ancient records had preserved their color, many undoubtedly being blacker than when used, but that the later records lost the jet-black appearance of the older. This, it is true, is not wholly due to the change of inks, for the use of quills, the soft surface of the old paper, the absence of blotting paper and the greater time spent in writing, were all conducive to a heavier deposit of ink; but evidence is ample that in comparatively recent years inks of poor quality came in use. Proof of this is given by an examination of the records in the state house. Up to about 1850 it was the custom in the office of the Secretary of the Commonwealth to use for engrossing the acts, inks made of a powder which was mixed in the office; and until that time the acts which are engrossed upon parchment show, with but few exceptions, no signs of fading. From 1850 for several years the writing in many cases is becoming indistinct, that upon an act in 1851, and upon two in 1855, having nearly disappeared. Since 1860, acts showing different intensity of color are found, but whether this is their original color or not cannot be determined.

That the fading can be attributed to the parchment, as some claim, is disproved by the fact that of the signatures upon the same act a few have faded while others have not. Upon an act approved January 4, 1845, the signature of the President of the Senate has nearly disappeared, that of the Speaker of the House is more legible, while that of the Governor, and the figure 4, which he evidently inserted, are jet black.

"The indexes in the volumes of archives in the office of the secretary, which were written about 1840, were evidently made with a different ink from that used for engrossing, and faded so badly that the important words had to be rewritten.

"In the office of the State Treasurer the records to about 1867 are very black and distinct, but the ink used during a few years following has faded.

"The records of births, marriages and deaths, in the registration volumes in the secretary's office, furnish an excellent illustration of the different qualities of the inks now used. These records are original returns made by the city and town clerks, and from 1842 to 1889 show instances of the use of inks which are now almost illegible. Here again the fault cannot be attributed to the paper, for endorsements made in the secretary's office upon the most faded returns at the time of their receipt are as black as when made.

"The volumes of copies of the old records of Lexington, made in 1853, have faded until they are quite indistinct.

"Some of the old inks, though retaining their black color have, from the presence of acid in the ink or paper, eaten through the paper as thoroughly as if the writing had been done with a sharp instrument. In part of one old volume of court records, the ink, while not injuring the paper or becoming illegible upon the face of the leaves, has gradually become legible upon the reverse, while the heavy paper has been impervious to the other inks used. * * * * * *

To ascertain what kind of inks were in use by the town clerks, I examined the registration volumes before referred to, and, as before stated, found many poor inks in use. In a few cases blue inks were used, and in two violet, which is, as a rule, if not always, a fugitive color. A number of the returns in these volumes of as recent date as 1875 were almost illegible, and three made in 1888 were nearly as indistinct.

"The more I looked into the subject, the more I became convinced that the whole subject of ink was one upon which the persons using it were comparatively ignorant. Consultation with experts satisfied me that good inks were being injured by improper treatment; that the custom of mixing inks and of adding water to them was unsafe;

and that among the inks reported as in use upon the records there were many manufactured for commercial uses which should not be used upon records, and which the manufacturers would say were not intended for record inks. I therefore sent to the manufacturers of the inks reported as in use by the recording officers, and to some others, the following letter and inquiries:

" 'The fading of much of the ink used in records of comparatively recent date, while as a rule the records of two hundred years ago are as legible as when written, establishes the fact that for permanent qualities much of the modern ink is inferior to the ancient, and that inks are used that are unfit for making a record which should stand for all time.

" 'I am led to believe that most ink in manufacturers make inks which are good for commercial and other uses where there is no desire for a permanent record, but which they would not recommend for use where the important object was the permanency of the record. One of the dangers to which our records are exposed can be obviated by the use of proper inks; and I desire to obtain the opinion of the leading manufacturers on the subject, that I may advise the recording officers of the State what are, and what are not, safe inks to use for records.

" 'I shall esteem it a favor, therefore, if you will answer the enclosed questions, and return them at your convenience. Your reply will be treated as confidential as far as names are concerned, except in the answer to question No. 5, and that will not be printed if you so request. Any general opinion which will aid the recording officers in their selection of ink or paper will be welcomed.

" '1. Do you consider it safe to use for a permanent record aniline inks?

" '2. Do you consider it safe to use for a record logwood inks?

" '3. Do your consider nut-gall and iron inks absolutely safe for a permanent record?

" '4. Do you consider carbon ink the only permanent ink?

" '5. What inks of your manufacture would you advise against using for a permanent record?

" '6. Do you advise generally against the inks known as writing fluids, when permanency is the first requisition?

" '7. Do you manufacture a writing fluid?

" '8. Do you consider it safe to add water to ink intended for permanent record, which has grown thick by exposure to the air?

" '9. Do you believe that the obliteration of ink is ever due to the chemicals left in the paper? (This question has been asked of the paper manufacturers also.)

" '10. Do you consider it safe to mix inks without knowing to what chemical group the inks so mixed belong?'

"Replies were received from twenty-two manufacturers. Several of the inks in the market, though bearing the name of certain persons, were found to be manufactured for them by manufacturers who had already answered the questions. Their replies were, therefore, not considered.

"To the first question, 'Do you consider it safe to use for a permanent record aniline inks!' the unanimous answer was decidedly no. Aniline black is absolutely permanent, but as it is not yet known how to render it soluble in water, it has not been much used in ink.

"To the inquiry in regard to logwood inks, nearly all answered no, and most of those who did not qualified their answers to such an extent as to imply distrust.

"Upon the question of the permanency of nut-gall and iron inks, the answers were more varied; one answering no, and four answering directly yes, the remaining answers being in brief that such inks were permanent if properly made.

"To the question, 'Do you consider carbon ink the only permanent ink?' the answers were varied and contradictory. Most of the manufacturers said a carbon ink could not be permanent, because carbon was insoluble; and some said that no chemical union could exist between carbon and the other ingredients in ink. Others claimed that carbon was the one permanent color, and cited the old Indian and Chinese inks which have stood for centuries as illustrations of its permanency. These statements were so widely different that I pursued the inquiry further, and found it was conceded that, if a process could be discovered by which carbon could be dissolved and made to retain its color, no known substance would make so permanent an ink; but that there was no such process, and in the inks now made the carbon was simply held in suspension in the ink without any chemical union; but I found also that improvement has been made, and that it is possible to combine the carbon with chemicals which will cause the carbon to embody itself. More than ordinary care should, however, be exercised in the purchase of carbon inks, for the lack of chemical union would cause a tendency to precipitate the carbon if the ink were improperly made.

"The replies to the inquiry, 'Do you advise generally against the inks known as writing fluids, when permanency is the first requisition?' were in a way the most unsatisfactory, and savored somewhat of advertising. One manufacturer made no fluid, and had no opinion to express. Most of the others made fluids. Nine advised generally against their use; four recommended them in preference to ink; and the others either advised generally against them, but recommended their own, or qualified the answer in such a way as to throw doubt on them.

"The argument in their favor seems to be that their fluidity makes them permeate the paper, and, in the change of color which usually takes place after using, a dyeing of the paper results. The objections are, that to obtain the fluidity body must be sacrificed, and there is not enough substance deposited upon the paper. The objections made by two manufacturers of fluids I give in their own words.

" 'We advise generally against the inks known simply as writing fluids—those not intended to yield a letter-press copy—because they are universally made, first, with as little solid matter as possible,—i. e. weak; second, with an excess of iron beyond that required to combine with the tannin, so as to develop all the color possible and flow with the greatest freedom. The combined writing and copying fluids, and the copying fluids on the other hand if properly made, may be justly recommended where permanency is the first requisition, particularly the older ones, which should be the most durable of all nut-gall and iron inks, because in them particularly concentration is aimed at, and the iron need not necessarily, and should not, be in excess of that required to combine with the tannin present. A steel pen during use injures, and often greatly, the durability of a writing ink by giving up iron to it.

" 'For your purpose, where extreme permanency is the first requisition, I should not advise the use of an ordinary writing fluid. Many manufacturers cannot obtain sufficient fluidity in their writing fluids without making their inks very dilute, and observing a particular method of manufacture which, although providing more attained color for a time, sacrifices the permanent quality of their color in a great measure. I should advise the use of an ink decidedly stronger.'

"The addition of water was almost universally condemned, for reasons stated later. As proof that this was not for the mercenary purpose of indirectly advising the use of more ink, some of the manufacturers said the ink should be kept in small- mouthed inkstands, and when not in use should be as tightly sealed as possible, to prevent evaporation.

"In reply to the inquiry as to whether chemicals left in the paper ever obliterated the ink, several of the manufacturers said they knew of such cases, and all were agreed that, if the chlorides used for bleaching the paper were not washed out, they would dangerously affect any ink. The practice of mixing inks was universally condemned.

"Permanency against the action of time is the quality sought for in this investigation, and it is claimed that better evidence as to that quality is furnished by the test of time than by any other; and manufacturers have shown or referred to specimens of writing made with their ink many years ago, as proof of its merit in this particular. If there was any surety that the standard of quality was always kept up in all of the oldest inks on the market, it would be safe to accept that test, but this may not be a fact; and, as has been stated, some of the recording officers believe that it is not.

Moreover, if only the old inks were to be accepted, it would be against the spirit of the age, which is to adopt the improvements which science makes possible; and manufacturers who at great cost of time and money have made improvements, would be deprived of the compensation which they deserve. The old inks were as a rule heavy, and had a tendency to settle; and the endeavor on the part of some manufacturers has been to preserve the permanency, and at the same time produce thinner inks which would be more agreeable to use.

"Improvements have been made in the direction of free-flowing inks, and these are fast becoming popular; and, while for correspondence and commercial uses they are undoubtedly sufficiently permanent, for records many of them are not, and it was with a view of preventing the use of these upon records that this investigation was made. No attention has been given to the permanency of the inks, as against their removal by acids.

"The use of proper ink is considered so important by the British government that the inks used in the public departments are obtained by public tender, in accordance with the conditions drawn up by the controller of H. M. stationery office, with the assistance of the chief chemist of the inland revenue department, to whom the inks supplied by the contractor are from time to time submitted for analysis. Suitable inks for the various uses are thus obtained, and their standard maintained. The last form of 'invitation to tender,' or 'proposal,' as we term it, is appended, as being instructive.

I cannot learn that the United States government uses any such care as the British government in the matter of ink, although the question has been a troublesome one in the departments.

"The State department issues no special rules for determining suitable inks, or requiring that particular inks shall be used. Proposals are asked for the lowest bids for the articles of stationery required, the last form of proposal asking for bids upon seven black inks, one crimson, and one writing fluid, which are named.

"With the market full of inks worthless for records, the only safety for our records seems to be in the establishment of a system similar to the English, which shall fix upon proper inks for various uses, which all recording officers shall be required to use.

"I believe that the recording officers will be glad to have the question of permanent inks decided for them, and to know whether inks which were in use many years ago, and have stood the test thus far, are maintained at their old standard. In the face of sharp competition among manufacturers, they fear they are not."

Mr. Swan, proceeding still further, secured the services of two of the most distinguished professors of chemistry in this country, Messrs. Markoe and Baird, and submitted to them in camera sixty-seven samples of different inks, known only by numbers, for chemical analysis; in a long and exhaustive report on the work they had set out to accomplish, and also with a dissertation on the chemistry of inks in general, they complete their report as follows:

"As a conclusion, since the great mass of inks on the market are not suitable for records, because of their lack of body and because of the quantity of unstable color which they contain, and because the few whose coloring matters are not objectionable are deficient in galls and iron, or both, we would strongly recommend that the State set its own standard for the composition of inks to be used in its offices and for its records, have the inks manufactured according to specifications sent out, and receive the manufactured products subject to chemical assay. In this way only can there be a uniformity in the inks used for the records throughout the State, and in no other way can a proper standard be maintained."

Mr. Swan comments on the report of his chemists, and calls attention to other tests made by himself:

"The conclusions at which I arrived were drawn, as stated, from manufacturers or recording officers, wholly independently of the chemists, but they will be found to coincide in many particulars with theirs. I did consult them in regard to the practicability of maintaining a State standard for record ink, which they have approved.

"The commendation by the chemists of some of the so-called writing fluids explains in a degree the variety of opinions advanced by the manufacturers in regard to the durability of fluids. Some of them will be seen to possess the qualities of ink, and the name fluid is evidently given to meet the commercial demand for fluids.

"Several persons, manufacturers among them, expressed greater confidence in tests of exposure of inks to the light and weather than to chemical analysis. I, therefore, as a dry test, placed on the inside of a window pane receiving a strong light, writing made under exactly the same conditions with each of sixty-seven inks, which remained there from March 13 to December 8. Similar writing was exposed to light and the weather from September 25 to December 8, and the result of the resistance of the inks in both tests is an almost exact confirmation of the report of the chemists, inks of the same class varying in their resistance according to their specific gravity or amount of added color.

"It may be safely said, therefore, that of sixty- seven inks of which I procured samples, all but seventeen are unsuitable for records, and among these the chemists say but one is fully up to the established scientific standard of quantity of iron sulphate. The reason is plain,—the demand for commercial inks is large, for record, small, and the supply has been to meet the demand."

The British government advertises for tenders each year, the requirements for black writing ink in 1889 reads:

"To be made of Best Galls, Sulphate of Iron, and Gum. The Sulphate of Iron not to exceed in quantity one-third of the weight of the Galls used, and the specific gravity of the matured Ink not to exceed 1045 degrees (distilled water being 1000 degrees)." That of Black Copying Ink "To be made of the above materials, but of a strength one fourth greater than the Writing Ink, and with the addition of Sugar or Glycerine. The specific gravity of the matured Ink not. to exceed 1085 degrees." And that of Blue-Black Writing Ink "To be made of finest Galls, Sulphate of Iron, Gum, Indigo, and Sulphuric Acid. The specific gravity of the Ink when matured not to exceed 1035 degrees."

Mr. Swan again remarks in his report of 1892:

"Many of the inks which should not be used upon records are free flowing and more agreeable to use than permanent inks, containing more body. As long as recording and copying is paid for by the page, and the object is to accomplish the most in the least time, these inks will be in popular use, and used, and blotted off the paper before they have much more than colored it, only to disappear eventually. The State should set a standard for a record ink; and, while our present system of keeping records and furnishing supplies will not allow that its use be required on all public records, as in England, it would seem practicable for the secretary of the Commonwealth to advertise for proposals for inks of a certain standard, which the manufacturers should be bound to maintain, and that these should be used in all the State offices. With a State standard ink adopted, its use by recording officers would soon follow."

In 1894 Mr. Swan's indefatigable efforts were crowned with success, the state of Massachusetts adopting his recommendations included in the following act:

"SECTION 1. No person having the care or custody of any book of record or registry in any of the departments or offices of the Commonwealth shall use or allow to be used upon such books any ink excepting such as is furnished by the secretary of the Commonwealth.

"SECTION 2. The secretary of the Commonwealth shall from time to time advertise for proposals to furnish the several departments and offices of the Commonwealth in which books of record or registry are kept with ink of a standard and upon conditions to be established by the secretary at such periods and in such quantities as may be required, and may contract for the same.

"SECTION 3. The ink so furnished shall be examined from time to time by a chemist to be designated by the secretary of the Commonwealth, and if at any time said ink shall be found to be inferior to the established standard the secretary shall have authority to cancel any contract made for furnishing said ink, and the quantity so found inferior shall not be paid for."

Professor Markoe, referred to before, was appointed "chemist" by the Secretary of the Commonwealth and prepared what he considered the best formula, for a standard ink, which was competed for by a number of ink manufacturers after proper advertisement, and a contract awarded. Mr. Swan says that this departure was received with favor by recording officers. No change was made in the formula until after the death of Professor Markoe in 1900, when Dr. Bennett F. Davenport of Boston was selected as his successor. He submitted a modified formula to be employed in the manufacture of an official or standard ink. It was adopted and such an ink is without exception now used by all recording officers of both Massachusetts and Connecticut.

In 1901 the United States treasury department adopted a similar ink except that it permitted the introduction into it of an unnamed blue coloring material.

Early in 1894 and during the legislative session of the state of New York, after consultation with General Palmer, the then secretary of state, I prepared a bill somewhat on the lines as laid down in the Massachusetts statute. The press all over the state at once took up the matter and urged that some such measure should be enacted into law. A New York City newspaper discussed it as follows:

"A bill is to be introduced in the legislature this week, probably to-morrow night, providing for an official ink to be used by every public officer throughout the State of New York in the writing of public documents and in making entries in the records.

"The official ink is for the purpose of making public records permanent and to guard against fraud by the alteration of the records. As the law stands at the present time in the state every official, whether municipal, county or state, is allowed to purchase and use for the records of his office whatever ink he may choose. The consequence is that there is no uniformity in public records throughout the state, and entries, transcripts and certificates are written with hundreds of various kinds of inks.

"The serious part of the business, however, is the evanescent character of some of the kinds now used, especially of the cheaper grades. These are the inks made from aniline and other dyes which are held in solution in water. Such inks are made from a fine, cheap powder, of which nigrosine is used in making black inks, eosine for red, and methylene for blue ink, and they cost only a few dimes a gallon to manufacture. The writing made with such inks quickly dries by the evaporation of the water, when it merely requires the application of a little soap and water to wash them out, leaving the paper absolutely clean, besides being fugitive.

"It is said that as a result of the present lack of system in this matter there are now public records of the city of New York in which the ink has entirely faded. These records have been made within the past forty years, and are now worthless because of the character of the inks originally used.

"In the Police department of this city a blue ink is often used which is made from prussian blue. A large portion of the entries in the books of the Police department are made with ink of this kind, and the warrants and other public documents with which the police have to do are similarly written.

"A little soap and water will wipe out this writing, so that the record can be easily altered at any time. The use of this ink in the Police department is said to date from the time of Tweed, which is significant of the original purpose for which it. was adopted.

"A permanent writing fluid such as it is now proposed to adopt throughout the state would not only secure uniformity in the character of the inks used, but it would also throw many obstacles in the way of altering the records.

"The present Secretary of State is heartily in accord with the proposed legislation. He was seen last week by Mr. David N. Carvalho, who has made a life study of the subject and who drew the bill and is pushing the reform.

"Mr. Carvalho said yesterday: 'This ink, whose use it is intended to secure in the making of public records in this state, is more costly than those made from aniline and other dyes, which fade and wash. In it the black particles are suspended in water by the addition of gum. This kind of ink has an affinity for oxygen, and hence it oxidizes and turns black. When unadulterated it only becomes blacker with the passage of time, and cannot be washed from the paper by the use of water.'

" 'I could show you,' continued Mr. Carvalho, 'public records of this city made within forty years which are entirely illegible and consequently worthless, because cheap inks were used in the writing. These include not only records of wills in the Surrogate's office, but entries and transfers of real estate which are likely to come up in the course of litigation at any time, thereby affecting the rights of many citizens.

" 'I can tell you at once upon seeing an old document the character of the ink that was used in the writing, and I have seen many old papers over a hundred years of age in which the writing was as clear as the day it was made, simply because a good writing ink was used. On the other hand writing made with cheap aniline ink may under certain circumstances fade out within a year, and in a book which is much handled is almost certain to be rubbed out in time.

" 'It has frequently happened that in the course of litigation, especially over real estate, that old records made with poor inks have been produced which the court refused to accept as evidence, thereby depriving some citizen of his rights. At the present time many officials in this state, in fact, the majority of them, are using these cheap and worthless inks and the records they are making will be of little or no value in a few years.

" 'It is to put a stop to this abuse that the present bill has been drawn up, and there is no argument which can be raised against it.' "

It appears that there was one, however, as the bill failed to pass for the stated reason that it came under the head of "class" legislation. The great state and city of New York with costly and magnificent depositories continue to place in them, for safe-keeping, valuable records and other ink-written instruments which will become illegible before the present century comes to an end.

Professor Lehner, a German chemist, in 1890 published a treatise "Die Tinten-Fabrikation," which has been translated and added to by Dr. Brannt, of Philadelphia, editor of "The Techno-Chemical Receipt-Book," who remarks:

"The lack of a recent treatise in the English language containing detailed descriptions of the raw materials and receipts for the preparation of Inks, and the apparent necessity, as shown by frequent inquiries, for such a volume, were the considerations which led to the preparation of The Manufacture of Ink."

This work compiles a great number of formulas, and rather favors the views of the chemist Dr. Bostock respecting the iron and gall inks. The book possesses value for reference purposes to the manufacturer.

Auguste Peret, author of "The Manufacture of Ink," 1891, has put together a lot of excellent material relative to ink-making and valuable for reference purposes.

The late Dr. William E. Hagan of Troy, New York, in 1894 issued his book, "Disputed Hand-writing." He devotes two chapters to the discussion of ancient and modern inks and their chemistry. He has been kind enough to quote the writer as the first to remove ink in open court with chemicals in order to determine the existence of pencil writing beneath the ink. The pencil being carbon was not affected thereby and with the subsequent restoration of the bleached ink by the use of the correct re-agent.

In the same year Dr. Persifor Frazer of Philadelphia published his "Manual of the Study of Documents." A few pages are given to the study of inks, and a part thereof is devoted to the researches of Carre, Hager, Baudrimont, Tarry, Chevallier and Lassaigne, to determine suspected forgeries. The chapter on "the sequence in crossed lines," where he indicates his method of determining which of two crossed ink lines was written first, is both original and a real contribution to science.

Alfred H. Allen, F. C. S., of England, perhaps the highest authority on the subject of tannins, dyes and coloring matters in his "Commercial Organic Analysis," revised and edited by Professor J. Merritt Mathews of Pennsylvania, edition of 1900, devotes eight pages to the subject of the "Examination of Ink Marks." He says:

"Ordinary writing ink was formerly always made from a decoction of galls, to which green vitriol was added. Of late, the composition of writing inks has become far less constant, aniline and other dyes being frequently employed, and other metallic salts substituted for the ferrous- sulphate formerly invariably used. The best black ink is a tanno-gallate of iron, obtained by adding an infusion of nutgalls to a solution of ferrous- sulphate (copperas)."

In 1897 the author in a paper read before the New York State Bar Association at Albany, entitled "A Plea for the Preservation of the Public Records," discussed the question of the stability of inks and their phenomena and took occasion to make recommendations as to their constitution and future methods of employment. A vote of thanks was adopted and the association referred the paper to the Committee on Law Reform, where no doubt it still slumbers.

CHAPTER XVI.

ENDURING INK.

ASCERTAINMENT OF A CORRECT INK FORMULA THE WORK OF OVER A CENTURY—CHARACTER OF THE EVIDENCE WHICH ESTABLISHES IT—THE INVESTIGATIONS OF THE AUTHOR IN THIS DIRECTION AND COMPARISON WITH THOSE OF COMMISSIONER SWAN—ELIMINATION OF THE "ADDED" COLORS AND THEIR ORIGIN— DISCUSSION OF THE RELATIVE MERITS OF LAMPBLACK, MADDER AND INDIGO—THE DURABLE VIRTUES OF INDIGO WHEN EMPLOYED ALONE—CAUSE OF THE BROWNING OF INKS—LONGEVITY OF INK DUE TO VEHICLE WHICH CARRIES IT—WHEN PERFECT INK WILL BE INVENTED.

TO ascertain the correct formula of a substantially permanent ink, as we have learned, has been the aim during a century or more, of able chemists, manufacturers and laymen. Their experiments and study of ancient and modern documents all point unerringly in the direction of an ink containing iron and galls.

Accumulated evidence may be said to establish itself in the light of investigation and experience and becomes more and more a certainty when considered, reviewed and discussed in connection with a chronological history of the "gall" inks since they came into semi-official and other uses centuries ago. Descriptions of MSS. containing ink writings hundreds of years old, many of them as legible as when first written, are silent witnesses whose testimony cannot be assailed. Such information when assembled together minimizes many of the conditions which have existed and interposed in preventing during the last four decades a general adoption or re-adoption of such a tanno-gallate of iron ink, the lasting qualities of which some of our forefathers estimated would, and as we know have stood the test of time.

Assuming this character of ink to have been employed in past centuries, the cause or causes for the differentiations in respect to color and durability become of paramount importance.

The investigations of the writer in this direction, while in some respects traveling the same road followed by others, diverged from them and has been more in the nature of a comparative analytical and microscopic examination of ancient with ancient and modern with modern documents in connection with numerous chemical experiments, the manufacture of hundreds of inks and the study of their time and other phenomena.

To accomplish this, ancient documents not written with "Indian" ink, but with those obviously containing combinations of iron and galls or other tannins, were selected and grouped into color families. They began with the fourteenth century, continuing well into the nineteenth, to the number of nearly four hundred, each of them of a different date and different year. Some of them were so pale and indistinct as to be illegible, others less so and by gradual steps they approached to a definite black; many of them as rich and deep in color as if they had been written not centuries ago but within a few years. Signatures on the same document represented different degrees of color, so that the question of the material on which the writing appeared affecting the appearance of the ink, was not a factor; but the difference in the inks used to make the signatures was the determining factor.

At this point it may be noted that the investigations conducted by Mr. Swan before referred to and those by the writer and the resultant observations of each were substantially alike. Many of the writer's, however, preceded those of Mr. Swan's, for during the years 1885 and 1886, having had the custody of part of the Archives of the City of New York there were many opportunities to study this subject which were taken advantage of, before and after which time frequent examinations were made of writings much more ancient than those pertaining to New York.

Assuming a second premise was to assert that the inks employed in the writing of these documents were "straight" or possessed some "added" pigment or color. Again, the vehicles to hold the particles or possibly preserving substances, might be factors.

All literature possible referring to ink formulas was examined to ascertain the names of materials recommended or formerly "added" to gall inks, because if the pristineness of the blacker inks was due to the added pigment it was a safe proposition that it was still existent in the ink, and that if it could be discovered part at least of the problem would be, simplified.

The "added" color compounds, excluding those of the aniline family which pertain to the more modern ink compositions, are of two classes: those possessing tannin and color-yielding materials and those containing only a color-yielding material. Many of the first class have been used in the manufacture of ink both with infusions of nut-galls or alone, while but very few of the second class have been used for either purpose. The decomposing action of light, oxygen and moisture on many of each class placed them beyond the purview of consideration, while the dates of the discovery and the fact of the small percentage of tannin contained in others permitted them also to be discarded. For instance: vanadium, which is fairly permanent, was discovered only in 1830; chanchi, the ink plant of New Granada discovered in the sixteenth century, possessing excellent lasting qualities, does not assimilate perfectly with other constituents used in the manufacture of ink, but is best when used alone; Berlin blue (prussian blue) is well spoken of, but was only discovered by accident in 1710 by Diesbach, a preparer of colors at Berlin; logwood, more used for this purpose than any other material, was first imported into Europe in the sixteenth century and causes a deterioration of the durable qualities of the tanno-gallate of iron; Brazil-wood and archil, and their allies, are exceedingly fugitive; bablah, the fruit of the acacia arabica, myrabolams, of Chinese growth, catechu, and sumac which though used in the time of Pliny, each contains a percentage of gallic acid too small to meet the requirements. Divi-divi, a South American product, came into use only at the end of the sixteenth century and has not stood the test of time.

This sifting process completely eliminated all but lampblack, madder and indigo in some form as a permanent "added" color pigment. Lampblack, which is we know forms the basis of "Indian" ink, is not soluble and requires a very heavy gummy vehicle to prevent its immediate precipitation, and while it could have been used in combination with tanno-gallate of iron as an ink, the fact that it was possible to chemically remove the ancient inks which remained black, was a sufficient demonstration that this carbon substance, which is not affected by chemicals, either as contained in the fluid ink or as dusted on after writing, could have formed no part of the ancient tanno-gallate of iron inks.

Madder is mentioned as of very ancient times and was cultivated in Europe as early as the tenth century; its addition to an iron and gall ink is said to be an invention of the year 1855; it is certain, however, that it was used for a like purpose as early as 1826, and a fair presumption that it was frequently employed in some form during the preceding four centuries. It has under certain conditions very lasting properties as the madder-dyed cloths found wrapped around Egyptian mummies demonstrates, but does not assist the tanno-gallate of iron to retain its black color; on the contrary it seems to lessen this quality.

That indigo for added color was employed by ink manufacturers in the eighteenth century is shown by the formulas appearing in the literature of that time. It was used alone as an ink long before, as well as contemporaneously with, those of the tanno-gallate of iron family. Its lasting properties are most remarkable if it be true that, used as a dye, there is still in existence specimens of it on cloth five thousand or more years old. The history of its use ALONE as an ink is difficult to ascertain back of a certain period; the writer has several specimens of it, one written in 1692 whose color is a green blue; another written about a century ago is believed to be as bright blue as the day it was placed on the paper; from 1810 to 1850 it was in common use particularly in hot climates where it was "home-made." Consequently if the old "gall" inks contained a lasting added color, indigo must have been the one, Dr. Stark whose investigations along this line for twenty-three years have already been cited has said that he preferred for his own use an ink composed of galls, sulphate of indigo and copperas (sulphate of iron); this means a tanno-gallate of iron ink with indigo for "added" color. Like formulas calling for different proportions of constituents both before and after his time in England and the continents of Europe and America are to be found in considerable number, proving that its use was more or less constant in this respect. To determine, then, whether or not the blacker specimens of the ancient writings contained indigo in any of its forms was most important, and the plan adopted most simple. Specimens of writing in ink of which the manufacturer's name was known as well as his formula and only thirty years old showed evidence of considerable "browning;" some of them when tested in juxtaposition with those of from fifty to one hundred years old which had turned completely brown, gave approximately the same results, and differentiated largely from the results obtained from jet black specimens of eighty to five hundred or more years of age. In a number of the browner ones indigo was found to be present while in many of the black ones it was not, demonstrating that the reason for the continuing blackness of the older inks is not due to an added color or pigment of any kind and furthermore that the "Stark" and corresponding ink formulas after the test of TIME did not retain their original blackness but deteriorated to a brown color; moreover, that their purpose as in the present day was to give an agreeable and immediate color result, a free-flowing ink, and to cheapen the cost of manufacture when compared with that of an unadulterated tanno-gallate of iron ink.

No disagreement being now possible as to the lasting color virtues of a properly proportioned tanno- gallate of iron ink WITHOUT an "added" color or pigment, there remained the sole question as to the vehicle utilized to hold this combination in suspension and whether or not it had to do with the continuing blackness of the older inks.

The answer must lie between the vegetable product known as gum and the animal product known as gelatine. The first disintegrates, quickly absorbs moisture and gradually disappears, while gelatine (isinglass) "contains under conditions 50% carbon, although its molecular formula has not yet been determined. It cannot be converted into vapor and does not form well-defined compounds with other bodies; it is insoluble in alcohol which precipitates it in flakes from its aqueous solution. It is also precipitated by tannin, which combines with it to form an insoluble non-putrescible compound. Gallic acid, however, does not precipitate it." (Bloxam.)

Possessing an undisturbed and complete history it was the very substance employed long before the discovery of gall ink, and is found present in the earliest specimens of the "Indian" inks which remain to us.

It must now be evident that there can be no material difference of opinions as to what has been so clearly and conclusively established, viz. that ink which contains a base of tanno-gallate of iron (without "added" color) is a permanent ink, and the length of its durability and continuing pristineness can be disturbed only by inferior quality of constituents, wrong methods of admixture and its future environment. Hence any black ink with this combination missing is of no practical value whatever either for record or commercial uses.

"Indian" ink, except for specific purposes, belongs to the great past and will so continue with its virtues unchallenged and proven, until some solvent is discovered for the carbon which forms nearly the whole of its composition, at which time THE perfect ink can be said to have been discovered.

CHAPTER XVII.

INK PHENOMENA.

CONDITION OF INK WHEN FIRST PLACED ON PAPER—ITS METAMORPHOSIS AND AFFINITIES—IGNORANCE OF THE FORGER AS TO ITS ORIGINAL ENVIRONMENT—TREATMENT OF OLD INK MARKS—HOW PAPER MAY DISCOLOR INK—THE USES OF ACID IN INK—VEHICLES TO HOLD INK PARTICLES AND PRESERVE THEM—INKS FIVE CENTURIES OLD DO PRESERVE THEIR GLOSS—SOME CAUSES OF INK DISINTEGRATION—WHEN INK BECOMES IRRESPONSIVE TO THE ELEMENTS— DEMONSTRATED TRUTHS ABOUT INK CONSTITUENTS AND COLOR PHENOMENA—NATURAL EVOLUTION OF AN INK MARK—LENGTH OF TIME REQUIRED TO BECOME BLACK—FIRST INDICATIONS OF AGE—DISAPPEARANCE OF INK QUALITIES—ARTIFICIAL AGING OF INK—TESTS FOR IT AND HOW TO CONFIRM THEM—BLEACHING AND REMOVAL OF INK FROM PAPER CRIMINALLY CONSIDERED— CHEMISTRY OF SUCH MARKS—THEIR RESTORATION— VARIATIONS IN METHODS WHICH CAN BE EMPLOYED.

ALL inks when first placed on paper are of course in a fluid state. Gradual evaporation of moisture causes a change not only in color but in the case of the iron and gall inks, in their chemical constitution, being immediately affected by their environment, whether due to the character of the paper on which they rest, the kind or condition of the pen used, or most important of all, the elements. Those who use the black inks and chemical writing fluids will have noticed these characteristics. The pale brown, blue or green as first written, and the gradual change after a short period to an approaching blackness, are reactions due largely to atmospheric conditions, the oxygen uniting with that for which it has affinity and instantly beginning with TIME to make its march, producing natural phenomena, which can be only superficially imitated but never exactly reproduced. When we further take into consideration that the forger cannot always know of the circumstances which surround the placing of original ink on paper and that be cannot manufacture the TIME which has already elapsed, it is not strange that attempted fraud can often be made evident and complete demonstrations given of the methods employed.

With the passage of time, the particles in some inks which are held together on the paper by gummy vehicles, commence to disintegrate and change from intense black to the brown color of iron rust, the "added" color which of itself is fugitive in character, soon departs; the vegetable astringent separating from the iron salt decays gradually and disappears and finally terminates in a mere stain or dust mark which can be blown off the paper. Sometimes, the written surface of such paper can be treated by carefully moistening it with a decoction of nut-galls or its equivalent in the presence of a weak acid, then if any iron be present, a measurable degree of restoration of color will ensue and remain for a short period.

Again, the discoloration of an iron ink may be due to the character of the paper; if of the cheaper grades and the bleaching compounds employed in their manufacture are not thoroughly washed out, then the ink not only begins to absorb oxygen from the atmosphere but the chlorine in the paper attacks it and the process of destruction is thereby hastened.

The introduction of acid into ink has two purposes, one to secure more limpidity, and the other to cause it to penetrate the paper and in this way bind together the constituent particles of both ink and paper. Most of the chemical writing fluids of this decade carry a superabundance of acid in their composition, which in time will burn through the paper and ultimately destroy it.

All tanno-gallate of iron inks require some vehicle to hold their particles in a state of suspension, otherwise there would be precipitation and such an ink could not be used. To meet this requirement a variety of gums are employed by manufacturers, gum acacia being the principal one. Its purpose is threefold—as before stated, to hold the ink particles in suspension—to prevent the ink from flowing too rapidly, and after drying WITHOUT blotting, to act as an envelope to encase the now fixed ink and prevent or interfere with its absorption of an excess of oxygen. The longer these latter conditions obtain the longer will the ink retain its pristineness, its durability and permanence. The "time proved" ink-written specimens of five hundred years or more ago which continue to retain their original intense black color and "glossy" appearance, do not, however, yield any evidence of the use of vegetable gums in their composition. Where such instances have been noticed the gloss is invariably missing. But, where ANY gloss is present, it was and is because of the employment of isinglass (fish-glue) as the vehicle to hold the ancient ink particles.

Hence the variations of color seen in ancient paper writings, as already stated, were due not only to possible imperfect admixtures of the component parts of the inks, but to the use of vegetable gums in their preparation. In the course of time these have been absorbed by moisture which hastened disintegration, causing a gradual disappearance of their original blackness and gloss and finally a return to the rusty color of oxidized iron.

It therefore follows, my observations and deductions being correct, the older a writing made with tanno-gallate of iron ink, where isinglass is the binder, and which has not been "blotted," the harder and more impervious and irresponsive it becomes to the action of the natural elements or of chemical reagents.

The truths demonstrated in this proposition cannot be denied. They fortify as certain that a properly proportioned mixture in water of an infusion of nut- galls or gallo-tannic acid and sulphate of iron, with isinglass as the vehicle to bold the particles in a state of suspension, if written with on good paper and allowed to dry without blotting, in a short time becomes encased or enveloped in such vehicle, which is thereby rendered substantially insoluble and absolutely prevents any extensive oxidation. Also, as a further consequent result, there is chemically created an unchangeable and continuing black color more permanent and durable than the substance on which it appears.

With a sample of standard commercial chemical writing fluid, write on "linen" paper without blotting it; in thirty hours, if exposed to the air and from three to five days if kept from it, the writing should have assumed a color bordering on black; it becomes black at the end of a month under any conditions, and so continues for a period of about five or six years, when if examined under a lens of the magnification of ten diameters, there will be a noticeable discoloration of the sides or pen tracks which slowly spreads during a continuing period of from ten to fifteen years, until the entire pen marks are of a rusty brown tint. A species of disintegration and decay is now progressing and when approximately forty years of age, has destroyed all ink qualities.

If, however, "chemical writing fluid" is first treated by exposure to the fumes of an ammoniacal gas, a "browning" of the ink occurs, not only of the pen tracks but of the entire ink mark. If examined now with a lens, the ink is found to be thin enough to permit the fibre of the paper to be seen through it, thus indicating artificial age. Furthermore, if a 20 per cent strength of hydrochloric acid be applied, the "added" color (usually a blue one) is restored to ITS original hue; alike experiment on "time" aged ink gives only the yellow brown tint of pure gall and iron combinations, the "added" color having departed caused by its fugitive characteristics. Again, if a solution of chlorinate of lime or soda be applied, the ink mark is instantly bleached, where in the case of honest old ink marks, it takes considerable time to even approximate a like result.

To confirm the chemical tests which may be employed in the determination of the artificial aging of ink marks, photographs made by permitting light to transmit through the paper and to interfere with its rays by filtering them through a "color" screen containing orange and some green, will indicate the presence of a fugitive substance in the ink, usually the "added" color employed in its manufacture.

The process of bleaching or "removal" of ink marks from paper is frequently employed in the attempted eradication of words or figures and the substitution of others on monetary instruments, commonly called "raising." Its purpose is usually a criminal one and some observations as to the modus operandi and its chemistry are not out of place here.

Ink marks made with a compound consisting of the combination of iron and an infusion of galls or its equivalent (a tanno-gallate of iron ink), as treated with certain chemicals, change from a compound with color to a chemical compound, with no color. Nothing has in fact been absolutely removed or eradicated, but it is a mere change of form, a sort of re-arrangement of the particles, the ingredients which formed the original color being still present, but in such a condition that they are invisible to the eye. A restoration of the invisible ink marks so that they can be observed, becomes possible by the use of chemical reagents and is the reverse of the one of erasure or bleaching, and changes the constituents again into a compound which has color from the one which had none. It does, not, however, reproduce the exact composition originally existing. Such a reagent simply goes to the basis of the material as first used, takes up what was left and reforms the particles sufficiently to make them abundantly recognizable. An apt illustration of these chemical changes of color is found in what is known as the phenolphtalein test solution, which is colored deep purplish-red by alkali hydrates or carbonates, and then by the addition of an acid rendered colorless, to be again reddened by an over- plus of the alkali and so on ad infinitum.

A popular material for the purpose of making chemical erasures is chlorinated lime or soda, which becomes more active by first touching the ink mark to be removed with a one half strength solution of acetic acid; this hastens the liberation of chlorine gas, THE active agent which causes the "bleaching" to take place. Hydrogen peroxide, also a bleaching compound, is less rapid in its action than chlorinate of soda; the same may be said of combinations of oxalic and sulphurous acids.

The most effective re-agent for the restoration of a chemically "bleached" iron ink mark is the sulphide or sulphuret of ammonia (it has several names). This penetrating chemical blackens metals or their salts, whether visible or not, if brought together. It must not be used by direct contact, the best and safest plan being to place a quantity in a small saucer, to be set on the floor of a closed box; to fasten to the box lid the specimen to be operated on; in this way the restoration is due to the fumes of the chemical and a possible danger of destruction of the specimen much lessened, especially if the marks are very light or delicate ones. The restoration of color under particular conditions may also be obtained by treatment with tannic acid, potassium ferro-cyanide (acidulated) or a weak solution of an infusion of galls.

CHAPTER XVIII.

INK CHEMISTRY.

SOME OBSERVATIONS AS TO CHEMICAL EXAMINATION OF INK MARKS BY ALLEN—ERASING OF INKS BY CHEMICAL MEANS—APPROVED CHEMICAL TESTS IN THE ASCERTAINMENT OF INK CONSTITUENTS.

A COMPILATION of the methods of Robertson, W. Thompson (Lord Kelvin), Irvine, Wislar, Hoffman and others, relative to the chemical examination of ink marks, is to be found in "Allen's Commercial Organic Analysis." Their experiments, however, date back many years ago, a few of them before the time of the use of the "anilines" for added color. The so-called "alizarin" ink referred to has now become obsolete. The following is the citation in part:

"In chemico-legal cases it is sometimes of importance to ascertain the nature of the ink used, to compare it with specimens of writing of known history, and to ascertain the relative ages of the writings. A minute inspection should first be made with a magnifying power of about 10 diameters, and any peculiarities of color, lustre, shade, etc., duly noted, and where lines cross each other which lie uppermost. The examination is often facilitated by moistening the paper with benzine or petroleum spirit, whereby it is rendered semi-transparent. The use of alcohol or water is inadmissible.

"Valuable information is often obtainable by treating writing or other ink-marks with reagents. Some inks are affected much more rapidly than others, though the rate of change depends greatly on the age of the writing. Normal oxalic acid (63 grammes per litre), or hydrochloric acid of corresponding strength, should be applied to a part of the ink marked with a feather or camel-hair brush (or the writing may be traced over with a quill pen), and the action observed by means of a lens, the reagent being allowed to dry on the paper. Recent writing (one or two days old) in gallic inks is changed by one application of oxalic acid to a light gray, or by hydrochloric acid to yellow. Older stains resist longer, in proportion to their age, and a deeper color remains. Log-wood ink marks are mostly reddened by oxalic acid, and alizarin marks become bluish, but aniline inks are unaffected. With hydrochloric acid, logwood ink marks turn reddish or reddish-gray, alizarin marks greenish, and aniline ink marks reddish or brownish-gray. The treatment with acid should be followed by exposure to ammonia vapors, or blotting paper wet with ammonia may be applied. Thus treated, marks in logwood ink turn dark violet or violet-black. The age of ink marks very greatly affects the rate of their fading when treated with dilute ammonia, the old marks being more refractory. The behavior of ink marks when treated with solution of bleaching powder is often characteristic, the older writings resisting longer; but unless the reagent be extremely dilute, writings of all ages are removed almost simultaneously. Hydrogen peroxide acts more slowly than bleaching solution, but gives more definite results. After bleaching the mark by either reagent, the iron of the ink remains mordanted on the paper, and the mark may be restored by treatment with a dilute solution of galls, tannic acid, or acidulated ferro- cyanide. The same reagents may be used for restoring writing which has been faded from age alone.

"When ink marks have been erased or discharged by chemical means, traces of the treatment are often recognizable. After effecting the erasure the spot is often rubbed over with a powdered alum or gum sandarac, or coated with gelatin or size. The bleaching agents most likely to have been used are oxalic, citric, or hydrochloric acid, bleaching powder solution, or acid sulphite of sodium. Moistened litmus paper will indicate the presence of a free acid, and in some cases treatment with ammonia fumes will restore the color. The presence of calcium, chlorides, or sulphates in the water in which the paper is soaked will afford some indication of bleaching powder or a sulphite having been used. Potassium ferro-cyanide will detect any iron remaining in the paper. Exposure to iodine vapor often affords evidence of chemical treatment, and other methods of examination readily suggest themselves."

M. Piesse, in the Scientific American, is authority for a method of removing ink, found on "patent" check paper:

"Alternately wash the paper with a camel's- hair brush dipped in a solution of cyanide of potassium and oxalic acid; then when the ink has disappeared wash the paper with pure water."

Inks of the tanno-gallate of iron family, whether containing "added" color or not, can be more or less "erased" by chlorinate of lime or soda, in the presence of a weak acid. These chemicals do not, however, materially affect the prussian blue inks, which require solutions of hydrate of potash or soda. Real indigo can be removed by chloroform, morphine or an aniline salt (indigo and aniline both owe their names to the same Portuguese source), which possess the rare property of dissolving pure indigo. Such combination, if refractory in the presence of permanganate of potash with sulphuric acid, must be followed by an application of sulphurous acid. In like manner, inks composed of by-products of coal tar, can be effectively treated, when irradicable with plain water or soap and water.

The erasure and removal of most inks from paper can be accomplished by the application of the chemicals heretofore enumerated. The requirements in this direction of some inks, however, though of rare occurrence, are to be met by the employment of other and particular reagents.

Many of the tests specified in the Allen citation to determine the character of ink constituents, if made alone are practically valueless, because the same behavior occurs with different materials employed in the admixture of ink. To avoid error in judgment the operator should verify if possible by confirmatory tests. Thus, in the one for logwood, sulphurous acid will cause a logwood ink mark to turn yellow; mercuric chloride, orange; tartar-emetic, red; and if the marks are faded ones, solutions of sulphate of iron or bichromate of potash will restore them respectively to a violet or blue-black color.

Prussian blue, aniline blue and indigo blue are to be tested as follows: Solution of chloride of lime, no change of color for prussian blue; decoloration or faint yellow for aniline blue or indigo. To discriminate between the two latter, test with solution of caustic soda, when decoloration or change of color will indicate aniline blue and permanence will indicate presence of indigo blue.

In the manufacture of the blue-black inks, a variety of violets have been and are still employed. Among them are aniline violet, iodine violet, madder, alkanet, orchil and logwood.

(a) Apply chloride of lime solution: 1. No change of color indicates alkanet. 2. Any change, one of the other five.

(b) Apply lemon juice: 1. The violet becomes brighter if it is one of the aniline violets, to be distinguished from each other by applying one part of hydrochloric acid to three parts of water, when it will become violet-blue, changing to red if it is common aniline-violet, but blue changing to a green hue and upon adding plain water to a lilac or pearl gray if it is iodine-violet (Hoffman's). It will also turn from red to yellow in lemon juice. To test for the other three violets: (a) Apply chloride of lime, to be followed by a solution of yellow prussiate of potash: absence of a blue coloration leaves orchil and logwood to be considered. To distinguish between them apply solution of hydrate of lime, whereby a change to gray, followed by complete decoloration indicates logwood, and a change to violet-blue, orchil.

The substances utilized with but few exceptions for red ink are the "eosins," possessing different names like erythrosine, as well as different hues. Antecedent to about thirty-five years ago, cochineal (known as "carmine"), madder, Brazil wood and saffron formed the basis of most of the red inks.

Make a soap solution adding a small quantity of ammonia, lemon juice, muriate of tin, all in water: 1. No change upon application indicates madder. 2. Any change, the presence of one of the three other reds: (a) thus a complete decoloration with a return of the color indicates saffron; (b) reappearance of the red color though weaker, aniline-red: (c) production of a yellowish red or light yellow color, cochineal or Brazil wood, to be distinguished from each other by the application of concentrated sulphuric acid, when Brazil wood will at once give a bright cherry-red, and cochineal a yellowish orange.

No yellow inks are in commercial use. Documents do, however, often contain yellow marks about which information is required as to their origin. As a rule they are iron rust, picric acid, turmeric, fustic, weld, Persian berries or quercitron. In order to recognize the different colors, the presence or absence of iron rust and picric acid must first be determined.

Apply a warm sample of a slightly acid solution of yellow prussiate of potash; iron rust will be indicated by a blue coloration.

Apply a weak solution of cyanide of potassium; picric acid will yield a blood-red coloration.

If picric acid and iron rust are both absent, apply a bit of ordinary wetted soap: 1. It turns reddish-brown and becomes yellow again with hydrochloric acid— turmeric; 2. It turns quite dark—fustic; 3. It is unaffected—weld, Persian berries or quercitron. To distinguish between these three, apply sulphuric acid, the color of weld will disappear, and of the others remaining apply tin-salt solution, when a change to orange indicates Persian berries, and no change or a very slight one, quercitron.

Inks containing also logwood, fustic, Brazil wood, or madder, were all of them more or less employed some years ago. Their color phenomena, following long periods of time, is much the same. Tests as prescribed in the accompanying table for such inks will serve to classify them preliminary to subsequent and more certain ones.

LOGWOOD. FUSTIC.

Concentrated Hydrochloric Acid Red-yellow Red
Dilute " " Reddish Yellow-Brown

Concentrated and dilute Nitric Acid Red Red-Yellow
" Sulphuric Acid . . Black Dark Purple
Dilute " " Red Brown Purple
Potassium Chromate . . . . Black
Stannous Chloride Violet Yellow
Tartaric Acid . . . . . Gray-Brown Yellow
Sulphate of Copper . . . . Dark Gray
Tannin . . . . . . Yellow-Red Yellow
Potash Dark Red Yellow
Potassium Permanganate Light-Brown Yellow
" Iodide . . . . . Red-Yellow
Pyrogallic Acid . . . . Yellow-Brown Yellow
Chrome-yellow . . . . . Dark Violet
Sodium (Salt) Violet Red
Sulphate of Iron Gray to Black

Alum . . . . Violet Red,Brown. Faint Red

BRAZIL WOOD. MADDER.

Concentrated Hydrochloric Acid Light Red Pale Yellow
Dilute " " Light Red Pale Yellow

Concentrated and dilute Nitric Acid Dark Purple Pale Yellow
" Sulphuric Acid . . Red Pale Yellow
Dilute " " Purple Pale Yellow
Potassium Chromate . . . . - -
Stannous Chloride Light Red Light Red
Tartaric Acid . . . . . Red Yellow Pale Yellow
Sulphate of Copper . . . . - -
Tannin . . . . . . No Change Pale Yellow
Potash Crimson Light Red
Potassium Permanganate - -
Iodide . . . . . - -
Pyrogallic Acid . . . . - -
Chrome-yellow . . . . . - -
Sodium (Salt) - Red
Sulphate of Iron Dark Violet -
Alum . . . . - Faint Red

CHAPTER XIX.

FRAUDULENT INK BACK GROUNDS.

DETECTION OF ALTERATIONS IN DOCUMENTS BY CHEMICAL TESTS WHICH APPLY SOLELY TO THE PAPER—ACCURACY OF RESULTS OBTAINED BY USE OF IODINE EXCELS THAT OF ALL OTHER CHEMICALS—IT APPLIES BEST TO LINEN PAPER—MODERN HARD PAPER DOES NOT GIVE COMPLETE INFORMATION—EFFECT OF IODINE ON MARKS MADE BY A STYLUS OR GLASS PEN.

FIFTY years ago and long before the employment of the fugitive "anilines" for ink uses, and "wood pulp" as a material for paper, two French chemists, Chevallier and Lassiagne, published in the Journal de Chimie Medical, an article "On the Means to be Employed for Detecting and Rendering Perceptible Fraudulent Alterations in Public and Private Documents," which as translated is valuable enough to quote in full:

"The numerous experiments which have been already tried at various times, have made known the processes which may frequently be put in practice for causing the reappearance of traces of writing effaced by chemical reactions, and for throwing light on the work of the guilty. But there are cases in which all the means proposed for this purpose fail, and then the criminal may escape justice from the want of conclusive material proofs. If, as has already been proved, it is not always possible to cause the reappearance of the effaced writing, for which written words have with a fraudulent intent been substituted, at least, as our experiments demonstrates, we may recognize, by some effects which are manifest on the surface of the altered paper, the places where the criminal act has been performed, circumscribe them by a simple chemical reaction visible to the least practiced eye, and even measure their extent. In a word, the visible alterations produced on a deed are susceptible, owing to the partial modifications which the surface of the paper has undergone, of being differently affected by certain chemical actions, and of being rendered visible. The following experiments, made in a judicial investigation, furnish us with the following facts:

"1st. The surface of paper sized in the ordinary way, or letter paper, no longer presents with certain reactions, the same uniformity where it has been either accidently moistened in several places by various liquids, or left in contact for a certain time with agents capable of removing or destroying the characters which have been traced on it with ink.

"2d. The application of a thin layer of gum, of starch, or farina, of gelatine, or fish-glue, with a view of sizing certain parts of the paper, or of causing certain bodies to adhere to it momentarily, is detected by an action similar to that which shows paper to have lately been wetted by the contact of liquids.

"3d. The heterogeneousness of the pulp of the papers, and the kind of size with which they are impregnated, lead to differences in the results which are observed with the same chemical reagents. We shall now examine each of these propositions, and describe the means which we have employed in endeavoring to solve questions of so high a degree of interest.

"1st. The homogeneousness of sized paper not partially altered by the contact of liquids (water, alcohol, salt-water, vinegar, saliva, tears, urine, acid salts, and alkaline salts) is demonstrated by the uniform coloration which this surface takes on being exposed, if not wholly, at least in various parts, to the action of the vapor of iodine disengaged at the ordinary temperature from a flask containing a portion of the metalloid. When the surface of paper not stained by any of the above mentioned liquids is exposed to the action of this vapor for three or four minutes in a room the temperature of which is about 60 degrees F., a uniform yellowish, or light-brownish yellow, coloration is noticed on the whole extent exposed to the vapor of iodine; in the contrary case, the surface which has been moistened, and afterwards dried in the open air, is perfectly distinguished by a different and well circumscribed tint. On the papers into which paste starch and resin have been introduced, the stains present such delicate reactions that we may sometimes distinguish by their color the portion of paper which has been moistened with alcohol from that which has been moistened with water. The stain produced by alcohol takes a bistre-yellow tint; that formed by water is colored of a more or less deep violet blue, the desiccation having been effected at the ordinary temperature. For the stains occasioned on these same papers by other aqueous liquids, the tint, apart from its intensity, resembles that of the stains of pure water. The feeble or dilute acids act like water on the surface of the same paper containing starch in its paste; but the concentrated mineral acids, by altering more or less the substances which enter into the composition of the latter, give test to the stains which present differences. We are always able to recognize by the action of the vapor of iodine the parts of the paper which have been put in contact with chemical agents, the energy of which has been arrested by washing in cold water. We are able, on several ancient deeds, written on stamped paper, and a few words of which had been removed by us with chemical agents, to recognize the places where their action was exerted, to see and to measure the extent which they occupied on the surface of the paper.

"The testing of a paper with the vapor of iodine will present this double advantage over the methods hitherto practiced for detecting falsifications in writings, that it points out at once the place in the paper in which any alteration may be suspected, and that, on the other hand, it enables us to act afterwards with the reagents proper for causing the reappearance of the traces of ink, when that is possible. If the means which we now propose cannot always make the former writing appear, they demonstrate the places where the alterations must have been made, when, however, the want of uniformity presented by the surface of the paper is not explained by any circumstance. This proof becomes, therefore, a weapon which the guilty person cannot avoid. But might not the presence of a stain, or several stains, developed by the vapor of iodine, in different parts of a public or private deed, give rise to a suspicion, where these stains have, perhaps, been occasioned by the spilling of some liquid on the surface of the paper? and would it not be rash and unjust to raise an accusation from such a fact? There would indeed be great temerity in drawing such a conclusion from a fortuitous circumstance; but the inference which may be drawn from the place occupied by these stains on the surface of the paper, from the more or less significant words found in those places, would not permit an accusation to be so lightly brought, where simple reasoning would be sufficient to destroy its basis. Besides, the subsequent reactions which would be made would certainly never revive words formerly written and effaced; whilst the latter effects may be often produced, more or less visibly, on those parts of the paper on which falsification has been practiced, figures or words being substituted for other figures or words.

"2d. The applications made to the surface of a sheet of paper, with a view of covering it again at certain parts with a fine layer of gum, gelatine, starch or flour paste, or in other places to cause other sheets of paper to adhere, may be recognized not only by the reflection of light falling upon the paper inclined at a certain degree of obliquity, and by the transmission of light through the paper, but also by the varying action which the vapor of iodine exerts on the surface which is not homogeneous. Papers containing starch and resin are more powerfully acted upon by this vapor than papers of a less complex composition. Both in the parts covered with starch, or paste flour, are colored in a few minutes of a violet blue; but with starched papers alone a more intense coloration is manifest on the places covered again with a thin layer of gum arabic, size or gelatine. By looking, then, on the surface of the paper, held somewhat obliquely to incidental light, we distinguish clearly, by their different aspects, the parts on which these various substances have been applied. The vapor of iodine, in condensing at the ordinary temperature on the surface of the papers to which any kind of size has been applied in various places, produces differences which are most commonly well recognized by the greater or less transparence of the paste of the paper.

3d. The heterogeneousness of the pulp of the various papers of commerce, and the nature of the size with which they are penetrated, cause differences, either in the coloration which the surface of these papers takes when exposed to the vapor of iodine, or in the tint which is manifested in the portions of the size deposited in certain portions of that surface; thus, papers with starched pulp generally turn brown, or blue, according to the amount of water that remains in their interstices; other papers turn yellow only under the influence of the vapor of iodine, and the parts which have received superficially a layer of another agglutinative body resist this action for a certain time, and are distinguished from the parts of the paper which are not covered with it."

My own investigations confirm to a great extent the value of these experiments and the accuracy of the deductions, in so far as they relate to "linen" paper; but they do not always obtain when made in connection with paper of inferior grades.

It is also true that dry paper is affected differently under the influence of the vapor of iodine, as would be paper which had been moistened and then dried; but the part which had been moist assumes the color of blue-violet, while unaltered paper assumes a yellow- brown color. Even when the paper thus treated is moistened all over with water, there will be a difference, for those parts which had been before moistened, will appear a dark violet-blue, while the other parts will show a plain blue coloration.

In cases where pencil writing has been removed with a soft rubber or fresh bread, the parts thus erased will assume, when subjected to iodine fumes, a brown color trending towards violet and much darker than the undisturbed portions of the paper. Lines impressed upon paper with a "stylus," a glass or ordinary dry pen, can be made visible by the fumes of iodine, the lines showing with a stronger coloration than the surrounding paper.

CHAPTER XX.

FUGITIVE INK.

SOME OBSERVATIONS ABOUT "ADDED" COLOR TO INK—INVENTION OF COAL TAR COLORS—CHRONOLOGICAL HISTORY OF THE "ANILINES" EMPLOYED AS INK—OTHER SUBSTANCES USED FOR THE SAME PURPOSE.

THE term "added color," as applied to ink, is the popular phraseology for a multitude of materials which have been more or less utilized for a period of centuries, in adulterating and coloring ink. In olden times they were introduced into ink with an honest belief that it would also improve and ensure its lasting qualities, but latterly more often to cheapen the cost of its manufacture. Reference has been made to a large variety of these substances used for this purpose and the story told of the effect of the test of time upon them as indicative of their supposed value. Attention has also been directed to the discovery during the nineteenth century of the colors which owe their origin to by-products of coal tar.

Generically these colors are classified as "anilines." They have worked a revolution in all the arts in which colors are used. Employed without a mordant, with few exceptions, they are measurably affected by both light, heat, moisture, or other changes and as made into inks are never permanent. Hence they should not be used for records, because if obliterated from any cause whatever, there are no known means to render them again legible.

The origin and history of the "anilines" are known. Viewed from an ink standpoint they are of vast interest. So extended in number are the "anilines" (they run into the thousands) that they include every shade of black and all possible tints or hues of the colors of the rainbow.

The chronological history of such of these artificial colors which appertain to ink or its manufacture is important as locating the dates of their invention and commercial use.

The first discovery of "aniline" is credited to Helot in 1750. In 1825 Faraday in rectifying naphtha discovered benzole, which by the action of strong nitric acid be converted into nitro-benzole; and this latter, when agitated with water, acetic acid and iron filings produced aniline. Unverdorben in 1826 discovered an analogous material in products obtained by the destructive distillation of indigo. Runge in 1834 claims to have detected it in coal tar and called it kyanol, which after oxidation became an insoluble black pigment and known as aniline black. It could not, however, be used as an ink. Zinan in 1840, experimenting along the same lines, produced another compound terming it benzidam. Fritsche in the same year by the distillation of indigo with caustic potash developed a product which he also called aniline, the name being derived from the Portuguese word anil, meaning indigo. Shortly afterwards A. W. Hoffman established the identity of these substances.

Aniline when pure is a colorless liquid, possessing a rather ammoniacal odor. It soon becomes yellow and yellow-brown under the influence of light and air. It does not affect litmus paper.

In 1856 Perkins accidentally discovered the violet dye called mauve, which acquired considerable commercial importance besides its utility for ink purposes.

Nicholson in 1862 succeeded in producing the first of the soluble blue anilines.

The discovery of induline, one of the modifications of aniline black, was made known in 1864.

Nigrosine, produced by the action of concentrated sulphuric acid on the insoluble indulines, was discovered in 1868.

The soluble indulines and nigrosines differentiate in appearance, the first a bronzy powder and the latter a black lustrous powder. When made into ink they possess about equal color values.

In 1870 the German chemists, Graebe and Liebermann, announced that they had succeeded in producing artificial alizarin,—the coloring matter of the madder root. Commercial value was not given to this discovery until it was put on the market in 1873, although it did not meet all the requirements.

Springmuhl in 1873 obtained an accessory product in the artificial manufacture of alizarin out of anthracene, from which a beautiful blue was made, superior in many respect to the aniline blues. It differed from aniline in having the same color in solution. Alkalis destroyed the color but acids restored it. The process was kept a secret for a long time. This product was originally sold as high as $1,500 for a single pound.

Caro, a German chemist, invented in 1874 the red color known as eosine, which was brought to this country in the following year and sold for $125 per pound. Its color is destroyed by acids.

Orchil or archil (the red color) was discovered in 1879. The commercial use of the so-called "orchil substitutes" (purples) began, however, in the years 1885 and 1887.

Artificial indigo, as the result of many years of experimenting, came into commercial use under the name of "indigo pure" only in 1897. It had previously been produced synthetically in a variety of ways, but the cost of the production was far above that of the natural product. Baeyer and Emmerling in 1870, Suida in 1878, Baeyer in

1878, Baeyer and Drewsen in 1882, and Heumann in 1890, can be said to have been the pioneers in the production of artificial indigo.

The intensity of some of the aniline colors may be indicated by the fact that a single grain of eosine in ten millions of water exhibits a definite rose-pink color.

It is asserted that in the last three years many improvements have been made in the permanent qualities of some of the soluble anilines, but no material which is soluble in plain water should ever be employed as an ink for record purposes.

Preceding the discovery of the "anilines," as already related, other substances had been employed for "added" color in the admixture of ink, principally madder, Brazil wood, indigo, and logwood.

Only a casual reference has heretofore been made to Brazil wood and logwood.

Brazil wood, also called peach wood, is imported from Brazil. Its employment as a dyestuff is known to be of great antiquity, antedating considerably the discovery of South America. Bancroft states, "The name 'Brazil' was given to the country on account of the extensive forests of the already well-known 'Brazil wood,' which was found by its Portuguese discoverers. The dyestuff thus gave its name to the country from which it was afterwards principally obtained. The word 'Brazil' appears to have been originally used to designate a bright red or flame color. Thus in a contract between the cities of Bologna and Ferrara, in 1194, the dyestuff kermez is referred to as grana de Brazile and Brazil wood, both dyestuffs at that time being obtained from India." For "added" color to ink and alone it was much used in the seventeenth and eighteenth centuries.

Logwood, employed more extensively for "added" color than any other color compound, was introduced into Europe by the Spaniards, A. D. 1502. In England it does not appear to have been much used until about 1575. In 1581 the Parliament prohibited its use "because the colours produced from it were of a fugacious character." Its use was legalized in 1673 by an act, the preamble of which reads, "The ingenious industry of modern times hath taught the dyers of England the art of fixing, the colours made of logwood, alias blackwood, so as that, by experience, they are found as lasting as the colours made with any sort of dyeing wood whatever." It is obtained principally from the Campeachy tree, which grows in the West Indies and South America.

The practical utility of logwood as the base for an ink was a discovery of Runge in 1848, who found that a dilute solution of its coloring matter, to which had been added a small quantity of neutral chromate of potassium, produced a deep black liquid which apparently remained clear and did not deposit any sediment. This composition became very popular on account of its cheapness and dark purple color. It is of a fugitive character, though, and has passed almost entirely out of commercial use.

CHAPTER XXI.

ANCIENT AND MODERN INK RECEIPTS.

"INDIAN" INK—SPANISH LICORICE—BITUMEN—CARBON FROM PETROLEUM—PROCESS TO OBTAIN GALLIC ACID—EFFECT OF SUGAR IN INK—DARK COLORED GALLS BEST FOR INK MAKING—SUBSTITUTES FOR GALLS—RELATIVE PROPORTIONS OF IRON AND GALLS—ANECDOTE OF PROFESSOR TRIALL—ESTIMATION OF SULPHATE OF COPPER—QUAINT INK RECIPE—RIBAUCOURT'S INK—HORSELEY'S INK—ELSNER'S INDELIBLE MARKING INK—BLACK INK FOR COMMON AND COPYING USES—COMMON BLACK INK—SHINING BLACK INK—PROCESS FOR "BEST" INK—INDELIBLE BLACK INK WITHOUT GALLS OR IRON—INK POWDER—STEEL PEN INK—SOME EARLY LITERATURE OF THE COAL TAR PRODUCTS—INK PLANT OF NEW GRANADA—"IMPERISHABLE" INK—FIRE-PROOF INK—"INERADICABLE" INK—EXCHEQUER INK—"PERMANENT" RED INK—SUBSTITUTE FOR "INDIAN" INK—TO PREVENT INK FREEZING—BACTERIA IN INK—GOLD AND OTHER INKS USED FOR ILLUMINATING.

INNUMERABLE receipts and directions for making inks of every kind, color and quality are to be found distributed in books more or less devoted to such subjects, in the encyclopaedias, chemistries, and other scientific publications. If assembled together they would occupy hundreds of pages. Those cited are exemplars indicating the trend of ideas belonging to different nations, epochs, and the diversity of materials. They can also be considered as object lessons which conclusively demonstrate the dissatisfaction always existing in respect to the constitution and modes of ink admixture. Many of them are curious and are reproduced without any amendments.

"Indian ink is a black pigment brought hither from China, which on being rubbed with water, dissolves; and forms a substance resembling ink; but of a consistence extremely well adapted to the working with a pencil-brush, on which account it is not only much used as a black colour in miniature painting; but is the black now generally made use of for all smaller drawings in chiaro obscuro (or where the effect is to be produced from light and shade only).

"The preparation of Indian ink, as well as of the other compositions used by the Chinese as paints, is not hitherto revealed on any good authority; but it appears clearly from experiments to be the coal of fish bones, or some other vegetable substance, mixed with isinglass size, or other size; and most probably, honey or sugar candy to prevent its cracking. A substance, therefore, much of the same nature, and applicable to the same purposes, may be formed in the following manner.

"Take of isinglass six ounces, reduce it to a size, by dissolving it over the fire in double its weight of water. Take then of Spanish liquorice one ounce; and dissolve it also in double its weight of water; and grind up with it an ounce of ivory black. Add this mixture to the size while hot; and stir the whole together till all the ingredients be thoroughly incorporated. Then evaporate away the water in baleno mariae, and cast the remaining composition into leaden molds greased; or make it up in any other form."

"The colour of this composition will be equally good with that of the Indian ink: the isinglass size, mixt with the colours, works with the pencil equally well with the Indian ink; and the Spanish liquorice will both render it easily dissolvable on the rubbing with water, to which the isinglass alone is somewhat reluctant; and also prevent its cracking and peeling off from the ground on which it is laid." * * * * * * *

There is found in small currents near the Baltick Sea, in the Dutchy of Prussia a certain coagulated bitumen, which, because it seems to be a juice of the earth is called succinum; and carabe, because it will attract straws; it is likewise called electrum, glessum, anthra citrina, vulgarly yellow amber.

"This bitumen being soft and viscous, several little animals, such as flies, and ants, do stick to it, and are buried in it.

"Amber is of different colours, such as white, yellow and black.

"The white is held in greatest esteem in physick, tho' it be opacous; when it is rubbed against anything, it is odoriferous, and it yields more volatile salt than the rest. The yellow, is transparent and pleasant to the eye, wherefore beads, necklaces, and other little conceits are made of it. It is also esteemed medicinal, and it yieldeth much oil.

"The black is of least use of all. (Sometimes used by the ancients in making ink.)

"Some do think that petroleum, or Oil of Peter, is a liquor drawn from amber, by the means of subterrenean fires, which make a distillation of it, and that jet, and coals are the remainders of this distillation.

"This opinion would have probability enough in it, if the places, from whence this sort of drogues does come, were not so far asunder the one from the other; f or petroleum is not commonly found but in Italy, in Sicily, and Provence. This oil distils through the clefts of rocks, and it is very likely to be the oil of some bitumen, which the subterranean fires have raised." * * * * * * *

There are various processes for obtaining gallic acid, one of which is to moisten the bruised galls and expose them for four or five weeks to a temperature of 80 degrees Fahr.; by which a mouldy paste is formed, which is pressed dry and then digested in boiling water, which after evaporation yields the acid, and mixed with the solution of green copperas, makes the, ink. A quicker process, however, is to put the bruised galls into a cylindrical copper of a depth equal to its diameter, and boil them in nine gallons of water—taking care to replace the water lost by evaporation. The decoction to be emptied into a tub, allowed to settle, and the clear liquid being drawn off, the lees are emptied into another tub to be drained. The green copperas must be separately dissolved in water, and then mixed with the decoction of the galls. A precipitate is then formed in the state of a fine black powder, the subsidence of which is prevented by the addition of the gum, which, separately dissolved in a small quantity of hot water, combines with the clear black liquid. Besides its effect in keeping the fine insoluble particles in suspension, the gum mucilage improves the body of the ink, prevents its spreading or sinking too much into the paper in writing, and also acts beneficially by forming a sort of compact varnish in it, which tends to preserve its colour, and shield it from the action of the air. If, however, too much mucilage is used, the ink flows badly from quill pens, and still more so from steel pens, which require a very limpid ink. The addition of sugar increases the fluidity of ink, and permits the quantity of gum to be increased over what it would bear without it; but, on the other hand, it causes it to dry more slowly, and besides it frequently passes into vinegar, when it acts injuriously on the pens. The dark- coloured galls, known as the blue Aleppo ones, are said by Ribaucourt, and others who have given much attention to the ingredients for ink-making to be the best for that purpose, and they are generally used by the best makers.

"From their high price, however, and that of galls generally, sumach, logwood, and even oak bark are too frequently substituted in the manufacture of inks, but it need scarcely be said always injuriously. Ink made according to the receipt given above is much more rich and powerful than many of those commonly made. To reduce it to their standard one half more water may be safely added; or even twenty gallons of tolerable ink may be made from the same weight of materials. Sumach and logwood admit of only about one-half or less of the green copperas that galls will take, to bring out the maximum amount of black colour. The colour of black ink gradually darkens in consequence of the peroxidation of the iron in it on exposure to the air, but it affords a more durable writing when used pale; its particles being then finer, penetrate the paper more intimately, and on its oxidation is mordanted into it. It is advisable so soon as the ink has acquired a moderately deep tint, to draw it off clear into bottles and cork them well.

"According to the most accurate experiments on the preparation of black writing inks, it appears that the proportion of the green copperas ought to be, and not to exceed, a third of the decoction of galls used; but the proportions used vary according to the practical experience of ink-makers, who have all receipts of their own, which they deem best, and, of course, keep secret. In the precipitate an excess of colouring matter, which is necessary for its durability, is preserved in it. The blue galls alone ought to be employed in making the best quality of black ink. Logwood is a useful. ingredient, because its colouring matter unites with the sulphate of iron and renders it not only of a very dark colour, but also less capable of change from the action of acids or of the atmosphere. Many attempts have been made by amateurs to make a good permanent black ink. A good story is told of Professor Traill. He had succeeded, after a long series of experiments, in producing an ink which he deemed to be in all respects A 1, and which resisted the action of all acids and alkalies alike. The pleased savant sent samples of it for trial to several banks and schools, where it gave general satisfaction; but, alas, an experimenting scribbler, thoughtlessly or otherwise, applied a simple test undreamt of by the Professor, and with a wet sponge completely washed off his 'indelible,' and thereby finished his career as an amateur ink-maker!" * * * * * * *

"Nicholson, in his Dictionary of Chemistry, an old but valuable work, says that Ribaucourt found vitriol of copper, in a certain proportion, to give depth and firmness to the colour of black ink; but, from whatever cause, this has not taken a place among the commonly-used ink-making ingredients— probably because it acts injuriously on steel pens." * * * * * * *

"A quart of rain Wate. 3 Ounces of Blue Knolly Gawalls. Bruise ym it must stand & be stirred 3 or 4 times in ym Day & then Strain out out all ye gawells all ten Days and 2 Ounces of Clear Gummary Beck & 1/2 an Ounce of Coperous 1/2 an Ounce of Rock Alum half an Ounce of Loafe sugar ye Bigness of a Hoarsel nut of Roman Vitterall Bray ym all small Before they be put in it must be stirred very well for ye space of two weeks.

"A receit forink.—1727

"William Satherwaite."

(The above receipt is a literal copy of the original, now in my possession. It purports to have been written with the mixture it specifies.) * * * * * * *

"M. de Champnor and M. F. Malepeyre, 1862, in their Mannel state that Ribaucourt's ink is one of the best then in use. The formula for its preparation is as follows:

Aleppo galls, in coarse powder, 8 ounces.
Logwood chips, 4 "
Sulphate of iron, 4 "
Powdered gum-arabic, 3 "
Sulphate of copper, 1 "
Crystallized sugar, 1 "

Boil the galls of logwood together in twelve pounds of water for an hour, or till half the water has been evaporated; strain the decoction through a hair sieve, and add the other ingredients; stir till the whole, especially the gum, be dissolved; and then leave at rest for twenty-four hours, when the ink is to be poured off into glass bottles and carefully corked. * * * * * * *

"Mr. J. Horsley gives the following receipt: Triturate in a mortar thirty-six grains of gallic acid with three and one-half ounces of strong decoction of logwood, put it into an eight ounce bottle, together with one ounce of strong ammonia. Next dissolve one ounce of sulphate of iron in half an ounce of distilled water by the aid of heat; mix the solutions together by a few minutes' agitation, when a good ink will be formed, perfectly clear, which will keep good any length of time without depositing, thickening, or growing mouldy, which latter quality is a great desideratum, as ink undergoing that change becomes worthless. It will not do to mix with ordinary ink, nor must greasy paper be used for writing on with it."— Chemical News (1862). * * * * * * *

"New Indelible Marking Ink.—Dr. Elsner gives the following as a stamping ink for goods before undergoing bleaching, or treating with acids or alkalis. It consists merely of one ounce of fine Chinese vermilion and one drachm of protosulphate of iron, well triturated with boiled oil varnish." * * * * * * * *

"Put Aleppo galls, well bruised, 4 1/2 oz. and logwood chipped, 1 oz. with 3 pints soft water, into a stoneware mug: slowly boil, until one quart remains: add, well powdered, the pure green crystals of sulphate of iron, 2 1/2 oz. blue vitriol or verdigris, (I think the latter better) 1/2 oz. gum arabic 2 oz. and brown sugar, 2 oz. Shake it occasionally a week after making: then after standing a day, decant and cork. To prevent moulding add a little brandy or alcohol.

"The common copperas will not answer so well as it has already absorbed oxygen." * * * * * * *

"Pour a gallon of boiling soft water on a pound of powdered galls, previously put into a proper vessel. Stop the month of the vessel, and set it in the sun in summer, or in winter where it may be warmed by any fire, and let it stand two or three days. Then add half a pound of green vitriol powdered, and having stirred the mixture well together with a wooden spatula, let it stand again for two or three days, repeating the stirring, when add further to it 5 ounces of gum arabic dissolved in a quart of boiling water, and lastly, 2 ounces of alum, after which let the ink be strained through a coarse linen cloth for use.

"Another. A good and durable ink may be made by the following directions: To 2 pints of water add 3 ounces of the dark coloured rough- skinned Aleppo galls in gross powder, and of rasped logwood, green vitriol, and gum arabic, each, 1 oz.

"This mixture is to be put into a convenient vessel, and well shaken four or five time a day, for ten or twelve days, at the end of which time it will be fit for use, though it will improve by remaining longer on the ingredients. Vinegar instead of water makes a deeper coloured ink; but its action on pens soon spoils them." * * * * * * * *

"Beat up well together in an iron mortar the following ingredients in a dry state; viz. 8 oz. of best blue gall-nuts, 4 oz. of copperas, or sulphate of iron, 2 oz. of clear gum arabic, and 3 pints of clear rain water.

"When properly powdered, put to the above; let the whole be shaken in a stone bottle three or four times a day, for seven days, and at the end of that time, pour the liquid off gently into another stone bottle, which place in an airy situation to prevent it from becoming foul or mothery. When used put the liquid into the ink-stand as required."

Take 6 quarts (beer measure) of clear water, soft or hard, and boil in it for about an hour 4 oz. of the best Campeachy logwood, chipped very thin across the grain, adding, from time to time, boiling water to supply in part the loss by evaporation; strain the liquor while hot, and suffer it to cool. If the liquor is then short of 5 quarts, make it equal to this quantity by the addition of cold water. After which let 1 lb. of bruised blue galls, or 20 oz. of the best common galls, be added. Let a paste be prepared by triturating 4 oz. of sulphate of iron (green vitriol) calcined to whiteness, and let half an ounce of acetite of copper (verdigris) be well incorporated together with the above decoction into a mass, throwing in also 3 oz. of coarse brown sugar and 6 oz. of gum Senegal, or Arabic. Put the materials into a stone bottle of such a size as to half fill it; let the mouth be left open, and shake the bottle well, twice or thrice a day. In about a fortnight it may be filled, and kept in well- stopped bottles for use. It requires to be protected from the frost, which would considerably injure it."

Infuse a pound of pomegranate peels, broken to a gross powder, for 24 hours in a gallon and a half of water, and afterwards boil the mixture till 1-3d of the fluid be wasted. Then add to it 1 lb. of Roman vitriol, and 4 oz. of gum arabic powdered, and continue the boiling till the vitriol and gum be dissolved, after which the ink must be strained through a coarse linen cloth, when it will be fit for use.

"This ink is somewhat more expensive, and yet not so good in hue as that made by the general method, but the colour which it has is not liable to vanish or fade in any length of time." * * * * * * * *

"Infuse a pound of galls powdered and 3 ounces of pomegranate peels, in a gallon of soft water for a week, in a gentle heat, and then strain off the fluid through a coarse linen cloth. Then add to it 8 oz. of vitriol dissolved in a quart of water, and let them remain for a day or two, preparing in the meantime a decoction of logwood, by boiling a pound of the chips in a gallon of water, till 1-3d be wasted, and then straining the remaining fluid while it is hot. Mix the decoction and the solution of galls and vitriol together, and add 5 oz. of gum arabic, and then evaporate the mixture over a common fire to about 2 quarts, when the remainder must be put into a vessel proper for that purpose, and reduced to dryness, by hanging the vessel in boiling water. The mass left, after the fluid has wholly exhaled, must be well powdered, and when wanted for use, may be converted into ink by the addition of water." * * * * * * * *

"Ten parts of logwood are to be exhausted with eighty of boiling water. To the solution one thousandth of its weight of yellow chromate of potash is to be added gradually. The liquid turns brown and at last blue-black. No gum is needed, and the ink is not removed by soaking in water. —Chemical Gazette, London (1850)." * * * * * * * *

"Shellac, 2 oz.; borax, 1 oz.; distilled or rain water, 18 oz. Boil the whole in a closely covered tin vessel, stirring it occasionally with a glass rod until the mixture has become homogeneous; filter when cold, and mix the fluid solution with an ounce of mucilage or gum arabic prepared by dissolving 1 oz. of gum in 2 oz. of water, and add pulverized indigo and lampblack ad libitum. Boil the whole again in a covered vessel, and stir the fluid well to effect the complete solution and admixture of the gum arabic. Stir it occasionally while it is cooling; and after it has remained undisturbed for two or three hours, that the excess of indigo and lamp- black may subside, bottle it for use. The above ink for documentary purposes is invaluable, being under all ordinary circumstances, indestructible. It is also particularly well adapted for the use of the laboratory. Five drops of creosote added to a pint of ordinary ink will effectually prevent its becoming mouldy." * * * * * * * *

"In November, 1854, Mr. Grace Calvert read a paper before the London Society of Arts in which he said that he hoped before long some valuable dyeing substances other than carbo-azotic acid would be prepared from coal tar.

"In another paper read before the same society in 1858 he said: 'This expectation has now been fulfilled. Messrs. Perkins and Church have obtained several blue coloring substances from the alkaloids of coal tar, and one from naphthalene.' Also that himself and Mr. Charles Lowe had succeeded in obtaining coal tar products yielding colors of a beautiful pink, red, violet, purple, and chocolate. (These were not soluble in water)." * * * * * * * *

"Among vegetable substances useful in the arts is one that has long been known in New Grenada under the name of the ink-plant, as furnishing a juice which can be used in writing without previous preparation. Characters traced with this substance have a reddish color at first, which turns to a deep black in a few hours. This juice is said to be really less liable to thicken than ordinary ink, and not to corrode steel pens. It resists the action of water, and is practically indelible. The plant is known as coryaria thymifolia." * * * * * * * *

"Desormeaux recommends that the sulphate of iron be calcined to whiteness; coarse brown sugar instead of sugar candy; 1/4 oz. acetate of copper, instead of one ounce of the sulphate, and a drop or two of creosote or essential oil of cloves to prevent moulding." (See Ribaucourt receipt, p. 194.) * * * * * * * *

"Mr. John Spiller communicated to the London Chemical News (1861) a paper on the employment of carbon as a means of permanent record. The imperishable nature of carbon, in its various forms of lamp-black, ivory-black, wood-charcoal, and graphite or black lead, holds out much greater promise of being usefully employed in the manufacture of a permanent writing material; since, for this substance, in its elementary condition and at ordinary temperatures, there exists no solvent nor chemical reagent capable of affecting its alteration.

"The suggestion relative to the mode of applying carbon to these purposes, which it is intended more particularly now to enunciate, depends on the fact of the separation of carbon from organic compounds rich in that element, sugar, gum, etc., by the combined operation of heat and of chemical reagents, such as sulphuric and phosphoric acids, which exert a decomposing action in the same direction; and by such means to effect the deposition of the carbon within the pores of the paper by a process of development to be performed after the fluid writing ink has been to a certain extent absorbed into its substance—a system of formation by which a considerable amount of resistance, both to chemical and external influences, appears to be secured. An ink of the following composition has been made the subject of experiment: "Concentrated sulphuric acid, deeply colored with indigo .... 1 fluid ounce. Water, ... 6 " " Loaf Sugar,.......................... 1 ounce, troy. Strong mucilage of gum-arabic 2 to 3 fluid ounces.

"Writing traced with a quill or gold pen dipped in this ink dries to a pale blue color; but if now a heated iron be passed over its surface, or the page of manuscript be held near a fire, the writing will quickly assume a jet black appearance, resulting from the carbonization of the sugar by a warm acid, and will have become so firmly engrafted into the substance of the paper as to oppose considerable difficulty to its removal or erasure by a knife. On account of the depth to which the written characters usually penetrate, the sheets of paper selected for use should be of the thickest make, and good white cartridge paper, or that known as 'cream laid,' preferred to such as are colored blue with ultramarine; for, in the latter case, a bleached halo is frequently perceptible around the outlines of the letters, indicating the partial destruction of the coloring matter by the lateral action of the acid.

"The writing produced in this manner seems indelible; it resists the action of "salts of lemon," and of oxalic, tartaric, and diluted hydrochloric acids, agents which render nearly illegible the traces of ordinary black writing ink; neither do alkaline solutions exert any appreciable action on the carbon ink. This material possesses, therefore, many advantageous qualities which would recommend its adoption in cases where the question of permanence is of paramount importance. But it must, on the other hand, be allowed that such an ink, in its present form, would but inefficiently fulfil many of the requirements necessary to bring it into common use. The peculiar method of development rendering the application of heat imperative, and that of a temperature somewhat above the boiling point of water, together with the circumstance that it will be found impossible with a thin sheet of paper to write on both sides, must certainly be counted among its more prominent disadvantages." * * * * * * * *

"Fire-proof ink for writing or printing on incombustible paper is made according to the following recipe: Graphite, finely ground, 22 drams; copal or other resinous gum, 12 grains; sulphate of iron, 2 drams; tincture of nutgalls, 2 drams; and sulphate of indigo, 8 drams. These substances are thoroughly mixed and boiled in water, and the ink thus obtained is said to be both fire- proof and insoluble in water. When any other color but black is desired, the graphite is replaced by an earthly mineral pigment of the desired color." * * * * * * * *

"Ineradicable Writing.—A French technical paper, specially devoted to the art and science of paper manufacture, states that any alterations or falsifications of writings in ordinary ink maybe rendered impossible by passing the paper upon which it is intended to write through a solution of one milligram (0.01543 English grain) of gallic acid in as much pure distilled water as will fill to a moderate depth an ordinary soup-plate. After the paper thus prepared has become thoroughly dry, it may be used as ordinary paper for writing, but any attempt made to alter, falsify, or change anything written thereon, will be left perfectly visible, and may thus be readily detected." * * * * * * * *

"Exchequer Ink.—To 40 pounds of galls, add 10 pounds of gum, 9 pounds of copperas, and 45 gallons of soft water. This ink will endure for centuries." * * * * * * * *

"Take of oil of lavender, 120 grains, of copal in powder, 17 grains, red sulphuret of mercury, 60 grains. The oil of lavender being dissipated with a gentle heat, a colour will be left on the paper surrounded with the copal; a substance insoluble in water, spirits, acids, or alkaline solutions.

"This composition possesses a permanent colour, and a MSS. written with it, may be exposed to the process commonly used for restoring the colour of printed books, without injury to the writing. In this manner interpolations with common ink may be removed." * * * * * * * *

Boil parchment slips or cuttings of glove leather, in water till it forms a size, which, when cool, becomes of the consistence of jelly, then, having blackened an earthern plate, by holding it over the flame of a candle, mix up with a camel hair pencil, the fine lampblack thus obtained, with some of the above size, while the plate is still warm. This black requires no grinding, and produces an ink of the same colour, which works as fregy with the pencil, and is as perfectly transparent as the best Indian ink." * * * * * * * *

"Instead of water use brandy, with the same ingredients which enter into the composition of any ink, and it will never freeze." * * * * * * * *

"Bacteria in Ink—According to experiments which have recently been completed at Berlin and Leipzig by the leading bacteriologists of Germany the ordinary inks literally teem with bacilla of a dangerous character, the bacteria taken therefrom sufficing to kill mice and rabbits inoculated therewith in the space of from one to three days." * * * * * * * *

"The most easy and neat method of forming letters of gold on paper, and for ornaments of writing is, by the gold ammoniac, as it was formerly called: the method of managing which is as follows:

"Take gum ammoniacum, and powder it; and then dissolve it in water previously impregnated with a little gum arabic, and some juice of garlic. The gum ammoniacum will not dissolve in water, so as to form a transparent fluid, but produces a milky appearance; from whence the mixture is called in medicine the lac ammoniacum. With the lac ammoniacum thus prepared, draw with a pencil, or write with a pen on paper, or vellum, the intended figure or letters of the gilding. Suffer the paper to dry; and then, or any time afterwards, breath on it till it be moistened; and immediately lay leaves of gold, or parts of leaves cut in the most advantageous manner to save the gold, over the parts drawn or written upon with the lac ammoniacum; and press them gently to the paper with a ball of cotton or soft leather. When the paper becomes dry, which a short time or gentle heat will soon effect, brush off, with a soft pencil, or rub off by a fine linen rag, the redundant gold which covered the parts between the lines of the drawing or writing; and the finest hair strokes of the pencil or pen, as well as the broader, will appear perfectly gilt."

It is usual to see in old manuscripts, that are highly ornamented, letters of gold which rise considerably from the surface of the paper or parchment containing them in the manner of embossed work;

and of these some are less shining, and others have a very high polish. The method of producing these letters is of two kinds; the one by friction on a proper body with a solid piece of gold: the other by leaf gold. The method of making these letters by means of solid gold is as follows:

"Take chrystal; and reduce it to powder. Temper it then with strong gum water, till it be of the consistence of paste; and with this form the letters; and, when they are dry, rub them with a piece of gold of good colour, as in the manner of polishing; and the letters will appear as if gilt with burnisht gold."

(Kunckel, in his fifty curious experiments, has given this receipt, but omitted to take the least notice of the manner these letters are to be formed, though the most difficult circumstance in the production of them.)

CHAPTER XXII.

INK INDUSTRY.

IMPORTANCE OF HONEST INK MANUFACTURE—ABSENCE OF INFORMATION AS TO NAMES OF MOST ANCIENT INK MAKERS,—WHERE TO LOOK FOR ANCIENT INK—THEIR PHENOMENAL IDENTITY—INK AND PAPER AS ASIATIC INVENTIONS ENTER EUROPE IN THE TWELFTH CENTURY—BOTH IN GENERAL USE IN THE FOURTEENTH CENTURY—MONKS AND SCRIBES AS THEIR OWN INK MANUFACTURERS—MODERN INDUSTRY OF INK BEGINS IN 1625—ITS GROWTH AND PRESENT SITUATION—THE GENERAL IGNORANCE OF THE SUBJECT—INK INDUSTRY IN THE EIGHTEENTH CENTURY—THE FIRST PIONEERS ABROAD AND THOSE AT HOME—OBSERVATIONS RESPECTING INK PHENOMENA OF THE PAST EIGHTY YEARS—WHAT SOME INK MAKERS SAY ABOUT IT—LITTLE DEMAND FOR PURE INKS—SOME SKETCHES OF THE LEADING INK MANUFACTURERS OF THE WORLD—ESTIMATION OF QUANTITY OF INK MADE IN THE UNITED STATES—THE "LIFE" OF A MARK MADE WITH ORDINARY WRITING FLUID—ESTIMATION OF MOST INKS BY PROFESSORS BAIRD AND MARKOE—FORMULA OF THE OFFICIAL INK OF THE STATE OF MASSACHUSETTS—VIEWS OF SOME PROMINENT INK MANUFACTURERS ABOUT SUCH INK—SOME COMMERCIAL NAMES BESTOWED ON DIFFERENT INKS—THE 200 OR MORE NAMES OF INK MANUFACTURERS OF THE NINETEENTH CENTURY.

THE consideration of the effect of the use of ink upon civilization from primitive times to the present, as we have seen, offers a most suggestive field and certifies to the importance of the manufacture of honest inks as necessary to the future enlightenment of society. That it has not been fully understood or even appreciated goes without saying; a proper generalization becomes possible only in the light of corroborative data and the experiences of the many.

History has not given us the names of ancient ink makers; but we can believe there must have been during a period of thousands of years a great many, and that the kinds and varieties of inks were without number. Those inks which remain to us are to be found only as written with on ancient MSS.; they are of but few kinds, and in composition and appearance preserve a phenomenal identity, though belonging to countries and epochs widely separated. This identity leads to the further conclusion that ink making must have been an industry at certain periods, overlooked by careful compounders who distributed their wares over a vast territory.

"Gall" ink and "linen" paper as already stated are Asiatic inventions. Both of them seem to have entered Europe by way of Arabia, "hand in hand" at the very end of the eleventh or beginning of the twelfth centuries and for the next two hundred years, notwithstanding the fact that chemistry was almost an unknown science and the secrets of the alchemists known only to the few, this combination gradually came into general vogue.

In the fourteenth century we find one or both of them more or less substituted for "Indian" ink, parchment, vellum and "cotton" paper. It was, however, the monks and scribes who manufactured for their own and assistants' use "gall" ink, just as they had been in the habit of preparing "Indian" ink when required, which so far as known was not always a commodity.

As an industry it can be said to have definitely begun when the French government recognized the necessity for one, A. D. 1625, by giving a contract for "a great quantity of 'gall ink' to Guyot," who for this reason seems to occupy the unique position of the father of the modern ink industry.

Ink manufacture as a growing industry heretofore and to a large extent at present, occupies a peculiarly anomalous situation. Other industries follow the law of evolution which may perhaps bear criticism; but the ink industry follows none, nor does it even pretend to possess any.

Thousands are engaged in its pursuit, few of whom understand either ink chemistry or ink phenomena. The consumer knows still less, and with blind confidence placidly accepts nondescript compounds labeled "Ink," whether purchased at depots or from "combined" itinerant manufacturing peddlers and with them write or sign documents which some day may disturb millions of property. And yet in a comparative sense it has outpaced all other industries.

With the commencement of the eighteenth century we find the industry settling in Dresden, Chemnitz, Amsterdam, Berlin, Elberfield and Cologne. Still later in London, Vienna, Paris, Edinburgh and Dublin, and in the first half of the nineteenth century in the United States, it had begun to make considerable progress.

Among the first pioneers of the later modern ink industry abroad, may be mentioned the names of Stephens, Arnold, Blackwood, Ribaucourt, Stark, Lewis, Runge, Leonhardi, Gafford, Bottger, Lipowitz, Geissler, Jahn, Van Moos, Ure, Schmidt, Haenle, Elsner, Bossin, Kindt, Trialle, Morrell, Cochrane, Antoine, Faber, Waterlous, Tarling, Hyde, Thacker, Mordan, Featherstone, Maurin, Triest and Draper.

In the period covered by the nineteenth century at home, the legitimate industry included over 300 ink makers. Those best known are Davids, Maynard and Noyes, Carter, Underwood, Stafford, Moore, Davis, Thomas, Sanford, Barnes, Morrell, Walkden, Lyons, Freeman, Murray, Todd, Bonney, Pomeroy, Worthington, Joy, Blair, Cross, Dunlap, Higgins, Paul, Anderson, Woodmansee, Delang, Allen, Stearns, Gobel, Wallach, Bartram, Ford and Harrison.

The ink phenomena included in the past eighty years has demonstrated a continuing retrogression in ink manufacture and a consequent deterioration of necessary ink qualities. When the attention of some ink makers are addressed to these sad facts, they attribute them, either to the demand of the public for an agreeable color and a free flowing ink, or to an inability to compete with inferior substitutes, which have flooded the market since the discovery of the coal tar colors; they have been compelled to depart from old and tried formulas, in the extravagant use (misuse) of the so-called "added" color.

An exceptional few of the older firms continue to catalogue unadulterated "gall" inks; but the demand for them except in localities where the law COMPELS their employment, is only little.

Interesting deductions can be made from the accompanying brief sketches of the leading ink manufacturers of the world.

The "Arnold" brand of inks possesses a worldwide reputation, although not always known by that name, beginning A. D. 1724 under the style of R. Ford, and continuing until 1772, when the firm name was changed to William Green & Co. In 1809 it became J. & J. Arnold, who were succeeded in 1814 by Pichard and John Arnold, the firm name by which it is known at the present day. This last named concern located at 59 Barbican, on the site of the old City Hall in London, and later moved to their present address, No. 155 Aldersgate street. The inks made by the "fathers" of the firm were "gall" inks WITHOUT "added" color. At the commencement of the nineteenth century we find them making tanno-gallate of iron inks to which were added extractive matter from logwood and other materials to form thick fluids for shipment to Brazil, India and the countries where brushes or reeds were used as writing instruments. For the more civilized portions of the world similar inks but of an increased fluidity were supplied, that the quill pens might be employed. The demands for still more fluid inks which would permit the use of steel pens, resulted in the modern blue-black chemical writing fluid, the "added" blue portion being indigo in some form. It was first put on the market in 1830. They manufacture over thirty varieties of ink, but only one real "gall" ink without "added" color.

In the early part of May, 1824, Thaddeus Davids started his ink factory at No. 222 William street, New York City. His first and best effort was a strictly pure tanno-gallate of iron ink, which he placed on the market in 1827 under the name of "Steel Pen Ink," guaranteed to write black and to possess "record" qualities. In 1833 he made innovations following the lines laid down by Arnold and also commenced the manufacture of a chemical writing fluid, with indigo for "added" color. Many more "added" colors were employed at different periods, like logwood and fustic, with the incorporation of sugar, glucose, etc. In the early fifties the cheap grades of logwood ink after the formula of Runge (1848) and which cost about four cents per gallon was marketed, principally for school purposes; it was never satisfactory, becoming thick and "color fading." Mr. Davids made many experiments with "alizarin" inks in the early sixties but did not consider them valuable enough to put on the market. In

1875 the firm introduced violet ink made from the aniline color of that name. Experimentations in 1878 with the insoluble aniline blacks and vanadium were unsuccessful; but the soluble aniline black (blue- black) known as nigrosine they used and still use in various combinations. During this long period their establishments have been in different locations. From No. 222 William street it was changed to Eighth street, with the office at No. 26 Cliff street. In 1854 the works were removed to New Rochelle, Westchester county, N. Y. In 1856 the firm name was Thaddeus Davids and Co., Mr. George Davids having been admitted as a partner and their warehouse and offices at this time were located at Nos. 127 and 129 William street, where a business of enormous proportions, which includes the manufacture of thirty-three inks and other products, is still carried on at the present day under the name and style of "Thaddeus Davids, Co." The old "Davids' Steel Pen Ink" continues to be manufactured from the original formula and is the only tanno-gallate of iron ink they make, WITHOUT "added" color.

The Paris house of "Antoine" as manufacturers of writing inks dates from 1840. They are best known as the makers of the French copying ink, of a violet- black color, made from logwood, which was first put on the market in 1853 under the name of Encres Japonaise. In 1860 an agency was established in New York City. They make a large variety of writing inks but do not offer for sale a tanno-gallate of iron ink without "added" color.

"Carter's" inks came into notoriety in 1861, by the introduction of a "combined writing and copying ink," of the gall and iron type and included "added " color. It was the first innovation of this character. At the end of the Civil War, John W. Carter of Boston, who had been an officer of the regular army, purchased an interest in the business, associating with himself Mr. J. P. Dinsmore of New York, the firm being known as Carter, Dinsmore & Co., Boston, Mass. In 1895 Mr. Carter died and Mr. Dinsmore retired from the business. The firm was then incorporated under the style of "The Carter's Ink Co." They do an immense business and make all kinds of ink. Of the logwoods, "Raven Black" is best known. When the state of Massachusetts in 1894 decided that recording officers must use a "gall" ink made after an official formula, they competed with other manufacturers for the privilege of supplying such an ink and won it. They do not offer for sale, however, "gall" ink WITHOUT added color. Their laboratories are magnificently equipped; the writer has had the pleasure of collaborating with several of their expert chemists.

The "Fabers," who date back to the year 1761, are known all over the world as lead pencil makers. They also manufacture many inks and have done so since 1881, when they built now factories at Noisy-le-Sac, near Paris. Blue-black and violet-black writing and copying inks of the class made by the "Antoines" are the principal kinds. They do not offer for sale, tanno-gallate of iron ink without "added" color. A branch house in New York City has remained since 1843.

"Stafford's" violet combined writing and copying ink was first placed on the New York market in 1869, though it was in 1858 that Mr. S. S. Stafford, the founder of the house, began the manufacture of inks, which he has continued to do to the present day. His chemical writing fluids are very popular, but he does not make a tanno-gallate of iron ink without "added" color, for the trade.

Charles M. Higgins of Brooklyn, N. Y., in 1880 commenced the manufacture of "carbon" inks for engrossing, architectural and engineering purposes, and has succeeded in producing an excellent liquid "Indian" ink, which will not lose its consistency if kept from the air. It can also be used as a writing ink, if thinned down with water. He does not make a tanno-gallate of iron ink without "added" color.

Maynard and Noyes, whose inks were much esteemed in this section for over fifty years, is no longer in business, as is the case with many others well known during the first half of the nineteenth century.

The enormous quantities of ink of every color, quality and description made in the United States almost surpasses belief. It is said that the output for home consumption alone exceeds twelve millions of gallons per annum, and for export three thousand gallons per annum.

It is very safe to affirm that less than 1/50 of 1 per cent of this quantity represents a tanno-gallate of iron ink WITHOUT "added" color. Most colored inks and "gall" ones which possess "added" color if placed on paper under ordinary conditions will not be visible a hundred years hence.

This statement of mine might be considered altogether paradoxical were it not for associated evidential facts, which by proving themselves have established its correctness and truth. To repeat one of them is to refer to the report of Professors Baird and Markoe, who examined for the state of Massachusetts all the commercial inks on the market at that time.

"As a conclusion, since the great mass of inks on the market are not suitable for records, because of their lack of body and because of the quantity of unstable color which they contain, and because the few whose coloring matters are not objectionable are deficient in gall and iron or both, we would strongly recommend that the State set its own standard for the composition of inks to be used in its offices and for its records."

An official ink modelled somewhat after the formula employed by the government of Great Britain was contracted for by the state of Massachusetts. It read as follows:

"Take of pure, dry tannic acid, 23.4 parts by weight.
of crystal gallic acid, 7.7 parts.
of ferrous sulphate, 30.0 parts.
of gum arabic, 10.0 parts.
of diluted hydrochloric acid, 25.0 parts.
of carbolic acid, 1.0 part.
of water, sufficient to make up the mixture
at the temperature of 60 degrees F.
to the volume of 1,000 parts by
weight of water."

Such an ink prepared after this receipt would be a strictly pure tanno-gallate of iron ink WITHOUT any "added" color whatever.

The estimation in which such an ink is held by the majority of the ink manufacturers is best illustrated by quoting from two of the most prominent ones, and thus enable the reader to draw his own conclusions.

"We do not make a tanno-gallate of iron ink without added color, and so far as we know, there is no such ink on the market, as it would be practically colorless and illegible." * * * * * * *

"There is no such ink (a tanno-gallate of iron ink without added color) manufactured by any ink- maker as far as I know. It is obsolete."

The commercial names bestowed on the multitude of different inks placed on the market by manufacturers during the last century are in the thousands. A few of them are cited as indicative of their variety, some of which are still sold under these names.

Kosmian Safety Fluid, Bablah Ink, Universal Jet Black, Treasury Ledger Fluid, Everlasting Black Ink, Raven-Black Ink, Nut-gall Ink, Pernambuco Ink, Blue Post Office Ink, Unchangeable Black, Document Safety Ink, Birmingham Copying Ink, Commercial Writing Fluid, Germania Ink, Horticultural Ink, Exchequer Ink, Chesnut Ink, Carbon Safety Ink, Vanadium Ink, Asiatic Ink, Terra-cotta Ink, Juglandin Ink, Persian Copying, Sambucin, Chrome Ink, Sloe Ink, Steel Pen Ink, Japanese Ink, English Office Ink, Catechu Ink, Chinese Blue Ink, Alizarin Ink, School Ink, Berlin Ink, Resin Ink, Water-glass Ink, Parisian Ink, Immutable Ink, Graphite Ink, Nigrilin Ink, Munich Ink, Electro-Chemical, Egyptian Black, "Koal" Black Ink, Ebony Black Ink, Zulu Black, Cobalt Black, Maroon Black, Aeilyton Copying, Dichroic, Congress Record, Registration, "Old English," etc.

The list of over 200 names, which follow, includes those of manufacturers of the best known foreign and domestic "black" inks and "chemical writing fluids" in use during the past century, as well as those of the present time.

Adriana
Allfield
Anderson
Antoine
Arnaudon

Arnold
Artus
Ballade
Ballande
Barnes
Bart
Bartram
Beaur
Behrens
Belmondi
Berzelius
Bizanger
Blackwood
Blair
Bolley
Bonney
Bossin
Boswell
Bottger
Boutenguy
Braconnot
Brande
Bufeu
Bufton
Bure
Carter
Caw
Cellier
Champion
Chaptal
Chevallier
Clarke
Close
Cochrane
Collin
Cooke
Coupier and Collins
Coxe
Crock

Cross
Darcet
Davids
Davis
Delunel
Delarve
Delang
Derheims
Dize
Draper
Druck
Duhalde
Dumas
Dumovlen
Dunand
Dunlap
Ellis
Eisner
Faber
Faucher
Faux
Featherstone
Fesneau
Fontenelle
Ford
Fourmentin
Freeman
Fuchs
Gaffard
Gastaldi
Geissler
Geoffroy
Gebel
Goold
Goupeir
Grasse
Green
Guesneville
Gullier

Guyon
Guyot
Haenles
Hager
Haldat
Hanle
Hare
Harrison
Hausman
Heeren
Henry
Herepath
Hevrant
Higgins
Hogy
Hunt
Hyde
Jahn
James
Joy
Karmarsch
Kasleteyer
Kindt
Klaproth
Kloen
Knaffl
Knecht
Lanaux
Lanet
Larenaudiere
Lemancy
Lenormand
Leonhardi
Lewis
Ley Kauf
Link
Lipowitz
Lorme
Luhring

Lyons
MacCullogh
Mackensic
Mathieu
Maurin
Maynard and Noyes
Melville
Mendes
Meremee
Merget
Minet
Moller
Moore
Mordan
Moser
Morrell
Mozard
Murray
Nash
Nissen
Ohme
Ott
Paul
Payen
Perry
Peltz
Petibeau
Platzer
Plissey
Pomeroy
Poncelet
Prollius
Proust
Pusher
Rapp
Reade
Redwood
Reid
Remigi

Reinmann
Rheinfeld
Ribaucourt
Ricker
Roder
Ruhr
Runge
Sanford
Schaffgotoch
Schleckum
Schmidt
Schoffern
Scott
Seldrake
Selmi
Simon
Souberin
Souirssean
Stafford
Stark
Stein
Stephens
Stevens
Syuckerbuyk
Swan
Tabuy
Tarling
Thacker
Thomas
Thumann
Todd
Tomkins
Trialle
Triest
Trommsdorff
Underwood
Vallet
Van Moos
Vogel

Wagner
Walkden
Wallach
Waterlous
Windsor and Newton
Winternitz
Woodmansee
Worthington

CHAPTER XXIII.

CHEMICO-LEGAL INK.

ESTIMATED VALUE OF SCIENTIFIC EVIDENCE AS HELD BY THE COURT OF APPEALS—NOW BEYOND THE PURVIEW OF CRITICISM—VERDICTS IN THE TRIALS OF CAUSES AFFECTED BY SUCH EVIDENCE—LENGTH OF TIME NECESSARY TO OVERCOME PREJUDICE AND IGNORANCE—WHERE OBJECTIONS TO SUCH EVIDENCE EMANATE—SOME OBSERVATIONS ABOUT SUCH EVIDENCE GENERALLY— WHEN PRECEDENT WAS MADE TO CHEMICALLY EXAMINE A COURT EXHIBIT BEFORE TRIAL—THE CONTROVERSY IN WHICH JUDGE RANSOM MADE THIS NEW DEPARTURE—CITATION OF THE CASE AND ITS OUTCOME— DECISION IN THE GORDON WILL CASE OBTAINED BY THE SCIENTIFIC EVIDENCE—COMPLETE STORY ABOUT IT—HISTORY OF THE DIMON WILL CASE AND HOW CHEMISTRY MADE IT POSSIBLE TO CONSIDER IT—OPINION OF JUDGE INGRAHAM—PEOPLE OF THE STATE OF NEW YORK V. CODY—THE ATTEMPT TO PROVE AN ALLEGED "GOULD" BIRTH CERTIFICATE GENUINE, FRUSTRATED BY CHEMICAL EVIDENCE—THE DEFENDANT CONVICTED—THE PEOPLE V. KELLAM—CHEMICAL EVIDENCE MAKES THE TRUTH KNOWN—THE HOLT WILL CASE AND THE EVIDENCE WHICH AFFECTED ITS RESULT—THE TIGHE WILL CASE—OPINION OF JUDGE FITZGERALD.

"The administration of justice profits by the progress of science, and its history shows it to have been almost the earliest in antagonism to popular delusion and superstition. The revelations of the microscope are constantly resorted to in protection of individual and public interests. . . . If they are relied upon as agencies for accurate mathematical results in mensuration and astronomy, there is no reason why they should be deemed unreliable in matters of evidence. Wherever what they disclose can aid or elucidate the just determination of legal controversies there can be no well- founded objection to resorting to them." Frank v. Chemical Nat. Bank, 37 Superior Court (J. & S.) 34, affirmed in Court of Appeals, 84 N. Y. 209.

THIS decision by a final court of adjudicature, expresses in no uncertain terms the now generally estimated value of evidence which science may reveal. The importance which that branch of it denominated "Chemico-legal ink" has attained and its utilization in many trials of causes both civil as well as criminal, places it beyond the purview of criticism or objection. With the introduction of a new class of inks in the last two decades, its scope has been much broadened.

Innumerable verdicts by juries wherever the system prevails, all over the world, the opinions of learned judges, whether presiding during a jury trial or sitting alone, more or less affected by this character of evidence, presents fairly the trend of the views of the public mind respecting it.

Constant experiment and successful demonstrations, covering a period of over fifty years, was necessary to overcome prevailing prejudices and ignorance.

The conditions to-day, which happily obtain, are that the objection to the introduction of such evidence finds its source usually in the side seeking to obscure and hide the truth or facts, while the honest litigant or innocent individual hastens to advocate its employment.

Another feature worthy of consideration is that persons who possess intimate knowledge of ink chem. istry and who might otherwise successfully perpetrate fraud if opportunity presented itself, refrain from making the attempt because of that very knowledge, which is sufficient also to teach them of the possible exposure of their efforts. Again, they and others are aware of the reliance placed on chemico-legal evidence as an aid to the cause of justice by courts and juries and this is an added reason why they hesitate to take chances. These propositions being true, they establish another one, viz: that most of the attempted frauds at the present time in this connection, are by the ignorant and those whose conceit does not permit them to believe that any one knows more than themselves.

Chemico-legal ink evidence as before stated has been employed in the trials of causes for many years; but it was not until the year 1889 that a precedent was established for the chemical examination of a suspected document preceding any trial. The honor of this departure from the ordinary modes of procedure belongs to the Hon. Rastus S. Ransom, who was surrogate of the county of New York at the time.

The matter in controversy was an alleged will executed in triplicate by one Thomas J. Monroe. Charges were made that the three wills were spurious, as they were facsimiles of each other. It was for the main purpose of determining the methods of their make-up that Judge Ransom rendered the opinion and made the order for its chemical examination which is cited in full:

Estate of Thomas J. Monroe.—"This is an application by the special guardian and contestant in this proceeding, which is now pending before the assistant, for leave to photograph the various papers which have been filed as the will of the deceased, and to compel the filing of two parts of one of said wills, which was executed in triplicate; likewise that the last paper be subjected to chemical tests for the purpose of disclosing the nature of the composition of the ink and the process or processes to which it has been subjected.

"Upon the oral argument the surrogate decided the applications first stated in favor of the petitioner, reserving only the question of his power to direct or permit the chemical tests. The special guardian on the oral argument stated that he was unable, to find any authority for the application.

"Consultation of the various sources of authority upon the subject of expert testimony and the various tests for the purpose of establishing or disproving handwriting has not resulted in the discovery of any authority for granting the application. It is apparent, however, from some of the cases that such an examination must have been permitted; for instance, in Fulton v. Hood (34th Penn. State Reports, 365), expert testimony was received in corroboration of positive evidence to prove that the whole of an instrument was written by the same hand, with the same ink, and at the same time. It is inconceivable how testimony of any value could be given as to the character of ink with which an instrument was written, unless it had been subjected to a chemical test. The writer of a valuable article in the eighteenth volume of the American Law Register, page 281 (R. U. Piper, an eminent expert of Chicago, Ill.), in commenting upon the rule as stated in the case of Fulton v. Hood (supra), very properly says:

" 'Microscopical and chemical tests may be competent to settle the question, but these should not be received as evidence, I think, unless the expert is able to show to the court and the jury the actual results of his examination, and also to explain his methods, so that they can be fully understood.'

"The writer of this article is also authority for the statement that in the French Courts every manipulation or experiment necessary to elucidate the truth in the case, even to the destruction of the document in question, is allowed, the Court, as a matter of precaution, being first supplied with a certified copy of the same.

"The most obvious argument to be urged against allowing a chemical test to be made on a will, and one that was suggested by the court on the argument of this motion, is that, inasmuch as the paper may be the subject of future controversy in this or some other tribunal, future litigants should not be prejudiced by any alteration or manipulation of the instrument. I do not think, however, that this objection is sound. Take an extreme case, of permitting a sufficient amount of the ink (which the affidavit of the expert shows to be but infinitesimal) for the purpose of chemical examination; the form of the letter would remain upon the paper; if not, the form and appearance of the entire signature might, as a preliminary precaution, be preserved by photography. The portion of the signature remaining would afford ample material for future experiments and investigations in subsequent proceedings wherein it might be deemed advisable to take that course.

"Because the subject matter of the controversy may be litigated hereafter should not deprive parties in the proceeding of any rights which they would otherwise have. They certainly are entitled to all rights in this proceeding that the parties to any future proceedings would have. Besides, all the parties whose presence would be necessary to an adjudication in, for example, an ejectment proceeding, are (or their privies are) parties here. It certainly cannot be that the law, seeking the truth, will not avail itself of this scientific method of ascertaining the genuineness of the instrument because of some problematical effect upon the rights or opportunities of parties to future litigations respecting the same instrument. The possibilities of litigation over a will are almost infinite, and if such a rule should obtain this important channel of investigation would be closed. Suppose the same objection were raised to the first action of ejectment which might be brought, it might then with the same force be urged that parties to some future ejectment suit would be prejudiced by a chemical test of the ink used in the will, and so on ad infinitum.

"By not availing itself of this method of ascertaining the truth as to the character of the ink, the Court deprives itself of a species of evidence which amounts to practical demonstration.

"I can see no reason why the application should not be granted."

The order in part reads:

"It is ordered and directed that Charles H. Beckett, the special guardian aforesaid, be and he hereby is allowed permission to photograph the aforesaid paper writings described in said order to show cause, viz., one of the two parts of a triplicate Will of Thomas J. Monroe, deceased, dated February 10th, 1873, which were filed in the office of the Surrogate of the City and County of New York on or about the 9th day of May, 1889, and also the contested Will herein dated March 27th and June 1st, 1888, and to have the said paper writing, bearing date March 22d and June 1st, 1888, subjected to such chemical test or tests as shall disclose the nature of the composition of the ink and, if possible, the process or processes to which it has been subjected, if any.

"And it is further ordered and directed that such chemical test be applied to the ink or writing fluid on said alleged Will to the following specified portion, or any part of such portions, viz."

Specifications in minute detail follow, calling attention to the words and spaces which are permitted to be chemically tested, and then continues:

"And it is further ordered and directed that the said paper writings shall be photographed before any chemical tests are applied thereto.

"And it is further ordered and directed that such photographing and chemical tests be performed by David N. Carvalho, Esq., a proper and suitable person, at the places above indicated respectively, between the 10th and the 20th days of June, 1889, inclusive, in the presence of the parties in interest or their attorneys, upon at least two days' notice to all parties herein or their attorneys.

"And it is further ordered and directed that in the event of destruction or breaking of the negatives after such paper writings have been photographed, the said special guardian, upon similar notice, shall have leave to re-photograph the said paper writings, at the same place and by the said David N. Carvalho, between the 10th and 20th days of June, 1889, inclusive. "(Signed) RASTUS S. RANSOM, "Surrogate."

On the 19th of June, 1889, pursuant to the order of the court, the alleged will referred to was first photographed, and later in that day such places as had been designated in the order were chemically treated, as part of a series of experiments. The results obtained briefly summarized were as, follows: The instrument which purported to be a holographic will of Thomas J. Monroe the experiments showed conclusively to be not the case, as neither pen nor ink in the body writing portion or in the decedent's signature had ever touched the paper; the date and names of the witnesses thereon were written, however, with pen and ink. Furthermore, the experiments demonstrated beyond question that exclusive of its date and names of witnesses, that it was what is commonly known as a transfer taken from a gelatine pad (hektograph), a method of duplicating popularly in vogue at that time. The deduced facts in the matter being that Thomas J. Monroe had written his will in an aniline purple ink, to which he had appended his name, leaving blank spaces to be filled in for the date, names of witnesses, etc., and had transferred the same to a hektograph, from which he had taken a number of duplicate facsimile copies, and at some other time had filled in the blank spaces by ordinary methods and to which, at his request, the names of the witnesses had been written with a pen and ink. In the trial which followed the surrogate declined to sustain the allegation of the proponents that the alleged signature was the original writing of Thomas J. Monroe, or indeed of any person. The will was not admitted to probate.

Experiments, both in open court or during its sessions in the testing of ink and paper, microscopically and chemically, are of frequent occurrence, and many contests involving enormous interests have been more or less decided as the result of them.

The contest of the alleged will of George P. Gordon, tried before the late Chancellor McGill of New Jersey in 1891, illustrates in a remarkable degree just how certain are the results of investigations of this character. The chancellor's decision, after listening to testimony for many weeks, was in effect to declare the will a forgery, largely because of the fact that the premise on which it rested was a so-called draft, from which it was sworn it had been copied. The ink on this draft it was proved could not have had an existence. until many years after the date of the forged will.

The decedent, who died in 1878, was the inventor of a famous printing press, and left a large fortune.

A will offered for probate soon after the death of Gordon was not probated, owing to the discovery that the witnesses had not signed it in each other's presence. The principal beneficiaries, however, under that will, the widow and daughter of Gordon, agreed to a division of the estate which was satisfactory to the other heirs at law, and the matter apparently was settled.

But a retired lawyer named Henry C. Adams began in 1879, a year after Gordon's death, to endeavor to obtain the assistance of some heirs at law in an enterprise which was finally ended only when Chancellor McGill's decision was rendered.

In 1868 Adams lived with his father and brothers on a farm, near Rahway, N. J., adjoining the Gordon place. The two men became well acquainted through their common interest in music. Adams called upon A. Sidney Doane, a nephew of Gordon, and told him that Gordon had made a will in 1868 which might be found or if lost, established by means of a draft of it which he (Adams) had retained. Mr. Doane refused to act upon this proposition. Then Adams presented the matter to Guthbert O. Gordon, a brother to George P. Gordon. He declined to consider the proposed search for a new will. Adams then wrote to Guthbert Gordon, Jr., cautioning him to say nothing to any one, but to come and see him. Guthbert Gordon, Jr., declined to accept Adams's invitation for a secret conference. Adams did not write or communicate with the widow or daughter of George P. Gordon, or with any of the officials or other persons who dealt with the estate. Finding that the heirs at law were satisfied with the arrangement of the estate under Gordon's daughter's management, he gave up his efforts at that time.

In 1890 Mary Agnes Gordon, the daughter, died in Paris, and remittances from her ceasing and her will not being satisfactory to those who had been receiving them from her, another contest was begun. This caused a renewal of Adams's activity. In 1890 he wrote to Messrs. Black & King, a firm of lawyers who represented the contestants of Mary Agnes Gordon's will. Adams's letter to the law firm contained this expression:

"If one of you will come over here on Sunday morning, bringing no brass band, fife or drums, I will tell you something worth knowing."

Mr. King visited Adams, who was then living at Orange, N. J., and was told by him that Mr. Gordon had executed a will in 1868 which he (Adams) had drawn at Gordon's instance, and that he had retained a corrected draft from which the will itself had been copied. He also told King that the original will after its execution had been left with his father, and that it must be at his father's homestead near Rahway, where he would try to find it. A few days later he wrote to Black & King that the will had been found, and the next day went with the lawyers to Rahway and identified the package found by his brother Edward Adams, who occupied the Rahway farm, as that which contained the will. The package, unopened, was taken to a safe deposit company and the original draft was deposited with the secretary of state. The alleged will, which Chancellor McGill pronounced a forgery when finally opened in the preliminary probate proceedings, was found to be a very long and complicated document, written on blue paper in black ink. The draft, which was on white paper, was also written in the main in black ink, but a copious quantity of red ink had been used in interlineations. The significant paragraph of the new will was a direction to his heirs to purchase, if the testator had not succeeded in doing so before his death, the Henry Adams farm for $32,000. Minute directions were given to insure the purchase, but no lower price than $32,000 was mentioned. Commenting upon this Chancellor McGill's remarks:

"It is also to be here noted that the Adams farm is now scarcely worth one-third the price for which it is directed to be purchased."

Continuing the court says:

"The only living person who professes to have had knowledge of this disputed paper prior to November, 1890, is Henry C. Adams. He most clearly and positively testified that he drew the disputed paper at the instance of Mr. Gordon. He produced a draft from which he said it was copied. . . . I have already stated that Mr. Adams testified most positively when the draft of the disputed paper was offered in evidence that it was the identical document from which the will of 1868 had been copied, and it is to be remembered that the interlineations in that draft are almost all made with red ink, and that Mr. Adams testified that those interlineations existed when the will was copied from the draft. With a view to testing the truth of this testimony the contestants submitted the draft to scientific experts, who pronounced the red ink to be a product of eosine, a substance invented by a German chemist named Caro in the year 1874, and after that time imported to this country. At first it was sold for $125 a pound, and was so expensive it could not be used commercially in the manufacture of ink. Afterwards the price was so greatly reduced that it became generally used in making red ink. It is distinguished by a peculiar bronze cast that is readily detected. It was recognized in the red ink interlineations in the draft of the disputed paper produced by Mr. Adams by a number of scientific gentlemen, among whom were some of the best known ink manufacturers in the country, and Mr. Carl Pickhardt, who first imported eosine. Upon further examination the witness, Adams, said he thought the draft produced to be the original until he saw the will on blue paper, and that then he was perplexed, but dismissed his doubt upon the suggestion of counsel, but afterward he thought upon the subject 'in the vigils of the night,' but by an unfortunate coincidence did not reach substantial doubt enough to correct his previous testimony until after the testimony concerning the character of the red ink he had used in interlining had been produced. . . . It is impossible to study this remarkable case at this point without grave doubts as to the truthfulness of Mr. Adams, and indeed as to the frankness with which the case was produced in court in behalf of the proponents."

As to Adams as a witness, the court finally says:

"And as I read the confused answers of Mr. Adams and note his apparent misapprehension of questions that would tend to involve him, and note the apparent failure of his theretofore wonderfully clear and exact memory of the most trivial and unimportant details, I am inclined to reject the whole story as a fabrication that has been punctured and fallen to pieces. . . . I find it to be impossible to rely upon the testimony of Henry C. Adams. Excluding it the will is not proved. . . .

"I will deny probate, revoking that which I have heretofore granted in common form." * * * * * * *

In the attempt made to prove the alleged last will and testament of Stephen C. Dimon, deceased, chemistry was the all-determining factor in the most important branch of the case. The peculiar features of this remarkable and unique case are best described by presenting them with a brief history of the entire matter.

In 1884 Stephen C. Dimon of the city of New York made and executed a will, choosing as legatee and executrix a Mrs. Martha Keery. The will he intrusted to the custody of his counsel. It appeared. that some time during the following year his attorney transferred this will from its resting place in a desk drawer to a new safe and recalled having seen its envelope a year later, but said he never saw the will thereafter.

In 1893 Mr. Dimon died. No will being produced, his brother took, out letters of administration. Whereupon Mrs. Martha Keery commenced a suit against the brother and the next of kin he represented, in an effort to obtain the dead man's estate. She based her claim solely on the LOST will, the contents of which were recalled in the trial by Mr. Dimon's former counsel, who was also one of the witnesses to the lost will. During the course of the trial in the Supreme Court, presided over by Justice George L. Ingraham, Mrs. Keery's attorney produced a mutilated document which from its reading indicated that it had once been a will, though not the "lost" one. But the names of the legatee, executrix, testator, names of witnesses and their addresses were completely obliterated. The written portions still undisturbed showed it to be in the handwriting of Stephen C. Dimon. Mrs. Keery's story was that after the death of Mr. Dimon in going over an old coat formerly worn by him, she had found it in a side pocket and had given it to her counsel just as it came into her hands.

Its condition showed it to be considerably pocket- worn. The obliterations referred to represented huge blots of black ink covering a lot of scratches and making it impossible to decipher the under writing. Defendant's Counsel immediately requested that the document be turned over to an expert, to see what could be done with it. The judge granted the motion and adjourned the case for several days to await results.

Counsel on both sides joined in the selection of myself. Three days were occupied in its decipherment. The will occupied two sides of a full sheet of legal cap. The original ink which was employed in the writing of the will was of pale gray color. The first obliterations were a series of pen and ink scratches and marks which destroyed the writing. Not satisfied with them the operator had with a saturated piece of blotting paper, brushed over the scratches and as that ink was of good quality every mark of writing had disappeared in the jumble and blots. It so happened that three inks had been employed. The original ink, the ink used for scratching and the one employed to do the blotting. The three inks were happily mixtures containing different constituents, and so by utilizing the reagent of one which did not affect the other, gradually the encrusted upper inks were removed and later the original writing appeared sufficiently plain not only to be read but to identify it. Photographs made before and after the chemical experiments, permitted court and counsel to make their own comparisons during the giving of the testimony about it.

It permitted also the finding of the two witnesses who lived outside of the city and to learn many details from them as to Mr. Dimon's conduct in the matter.

The restored will showed that Mrs. Keery at its date (1891) was still in his mind, and its destruction by himself—that he had changed his mind.

Justice Ingraham completes his opinion in deciding the case as follows:

"In this case, however, the long time that elapsed between the time of the delivery of the will to Mr. Morgan and the death of the testator, the absence of my satisfactory proof of the existence of the will from the time it was delivered to Mr. Morgan to the time of the testator's death, and the fact that the testator made another will, making substantially the same disposition of the property, which he subsequently destroyed, all tend to cast a doubt upon the fact that the will was in existence at the time of the testator's death, and there is positively no evidence that it was ever fraudulently destroyed.

"I do not think the court is justified in diverting a large sum of money from those legally entitled to it, by allowing, a lost will to be proved, except upon the clearest and most satisfactory evidence of the existence of the will at the time of the testator's death. And the testimony in this case falls short of what I consider necessary to establish such a will.

"There should be, therefore, judgment for the defendants with costs." * * * * * * *

A case of considerable interest was tried before Hon. Clifford D. Gregory in the month of March, 1899, in the city of Albany, New York. It was entitled the "People of the State of New York against Margaret E. Cody," as charged with the crime of blackmail, in the sending of a letter to Mr. George J. Gould, in which she threatened to divulge certain information which she claimed to possess about his dead father, Jay Gould. The character of this information was such that if true it meant that Jay Gould and his wife had lived in bigamous relations during a great number of years preceding their death and hence also affected the legitimacy of the entire Gould family. Mrs. Cody asserted that Jay Gould was married to a Mrs.

Angel some time in 1853, and that as a result of that "lawful" marriage she gave birth to a daughter, a Mrs. Pierce, who was still alive and living somewhere in the west. As Mrs. Cody offered to sell or secrete the information which she said she possessed for a consideration, Mr. George J. Gould and his sister, Miss Helen Gould, instantly determined that it could be nothing else than a clear case of an attempt at blackmail, which falsely impugned the reputations of their dead parents. They instituted criminal proceedings against Mrs. Cody, charging that Mrs. Cody when she wrote the letter well knew that her claim that his father had been married to Mrs. Angel and that Mrs. Pierce was their daughter, was absolutely false. Two trials followed, the first in 1898 in which the jury disagreed, and a second one in 1899 which lasted over a week. It was in the second trial that chemical tests on a certain entry in a church record in the presence of the jury were made, which showed conclusively that ancient writing of another character than that which had been substituted was still existent beneath the writing which was apparent to the naked eye.

The following are excerpts of the judge's charge to the jury:

"I wish to invite your attention, for a few moments, to the baptismal certificate. You have had produced here before you the original baptismal record of the church at Cooperville. It has been substantially admitted, in the arguments of this case, that there has been a change made in this certificate. I do not think that the District Attorney claims that there is any evidence that Mrs. Cody herself changed this record; there is no claim, as I understand it, made by the prosecuting officer that she went there and obtained this book, and with her own hand changed this record; but he asks you to infer and find from the evidence that has been given, that she was a party to this change, that she was privy to this change, and that knowing that fact she had guilty knowledge when she wrote the letter upon which the indictment is based.

"You will remember that Mr. Carvalho, the expert in handwriting, was placed upon the stand; and he has testified in your presence as to his qualifications in determining disputed handwritings, and what his experience has been during a long series of years. He tells you that he has examined this record, and that there is no question but some of the words have been erased and others substituted in their places. He tells you that the words 'Jay Goulds' were not the original words in the certificate, or if they were, the present 'Jay Goulds,' as they appear in the certificate, have been forged; that the words 'Mary S. Brown,' the 'sex mois,' the French words for six months, and other changes which he has described to you are forgeries.

"I shall submit to you, as a question of fact, whether or not Mrs. Cody had any knowledge or took any part, or authorized or connived at any of the changes made in this certificate. I do not say that she did; I leave it to you to say, from the evidence in this case, whether your minds are convinced that she had any part or parcel, or undertook in any way to accomplish the changes which have been made in this baptismal record. And if you find as matter of fact that she had such knowledge at the time this letter was written; if you find as matter of fact she had this information given to her by Mrs. Angel, then I leave it to you to say whether she had such knowledge, such guilty knowledge, as should prevent her, if acting honestly, from writing a letter such as has been described here and contained in the indictment."

The jury brought in a verdict of guilty.

In the trial of the People v. David L. Kellam (1895), who was charged with altering the dates of three notes for $6,000 each, the contention of the prosecution was that the dates of the notes had been changed by chemicals, and with the consent of the defense a reagent was applied to the suspected places and the original dates restored. The verdict of the jury was guilty.

In the Holt Will case, tried in Washington, D. C., in the month of June, 1896, great stress was laid on the fact of the difference in the admixture of inks found on letters contemporaneous with the date of the will, and it was asserted also that the ink with which the will was written was not in existence at the time it was alleged to have been made, June 14, 1873, and probably not earlier than ten years later. Furthermore, that it was a habit of Judge Holt up to the time of his death, which habit was illustrated in his writings and correspondence to "sand" his writing. The jury decided the will was a forgery.

Another famous case in which the scientific testimony about ink and pencil writing must have assisted the court in arriving at a conclusion was in the trial of the famous Tighe will contest, tried before Hon. Frank T. Fitzgerald, one of the present surrogates of the county of New York. The story of this case is incorporated in the opinion which is cited in part:

"Hon. Frank T. Fitzgerald, Surrogate of the county of New York:

"That Richard Tighe died on the 6th day of May, 1896, at No. 32 Union Square, in the city and county of New York, where he had lived for fifty years prior to his death, and was at the time of his death over ninety years.

"That the testator, on or about the 27th day of March, 1884, in the presence of the attesting witnesses, duly signed the instrument in writing, and duly published and declared the same to be his last will and testament, and requested said witnesses to witness the same, and pursuant to such request said attesting witnesses did subscribe said will as attesting witnesses. That at the time said Richard Tighe so signed, published and declared the said instrument to be his last will and testament, the said Richard Tighe was in all respects competent to execute the same, and was not under any restraint or undue influence. That the said instrument, so signed, published and declared by testator was and consisted of the identical sheets of paper and the identical writing now appearing upon the same as to all except pencil writing; the testator did not publish or declare the marks, words or figures written in or upon said instrument in pencil to be a part of his last will and testament, and it is not found that such marks, words or figures were upon said instrument at the time when said instrument was so published and declared to be the last will and testament of the testator. That the said last will and testament is written consecutively upon two sheets of legal cap paper.

"That the said last will and testament was originally prepared with blank spaces left for the insertion of the numbers of shares intended to be bequeathed and devised to the various beneficiaries named therein, and as so prepared was in the hand-writing of Caroline S. Tighe, the wife of testator, and that at some subsequent time and before the execution of the said instrument by the said Richard

Tighe, the blank spaces hereinafter referred to as filled in in ink, were filled in by or under the direction of the testator. Upon said instrument as offered for probate there appears in the blanks originally left thereon, in some instances, pencil writings superimposed over other pencil writings, which have been either wholly or partially erased, and in other instances ink writing different from the body of the instrument in the material employed, appearing over pencil writings wholly or partially obliterated. . .

"That the said words written in ink filling such blanks as aforesaid expressed the final determination of the testator with regard to the beneficiaries to whom the same applied; and that the words and figures written in pencil filling such blanks as aforesaid were written only deliberately and tentatively and that as to those words and figures the testator had not at the time when he executed, published or declared said instrument to be his last will and testament determined as to whom or in what proportions he would give the several shares of his estate and property covered by said words and figures, but the testator attempted and intended to reserve to himself the power of making disposition of said shares thereafter, and intended the final disposition thereof to be in ink writing. . . ."

CHAPTER XXIV.

CHEMICO-LEGAL INK (CONTINUED).

FAMOUS CASE OF CRITTEN V. CHEMICAL NATIONAL BANK—STORY OF THE CASE INCLUDED IN THE OPINION OF THE COURT OF APPEALS AS WRITTEN BY JUSTICE EDGAR M. CULLEN—THE PINKERTON CASE OF "BECKER"—STORY OF HOW HE SECURED $20,000 THROUGH THE ALTERATION OF A $12 CHECK—BECKER'S COMMENTS ABOUT HIMSELF—A CRITICISM OF BECKER AND HIS WORK—NAMES OF SOME CASES IN WHICH CHEMICAL EVIDENCE WAS PRESENTED TO COURTS AND JURIES.

THE books contain no clearer or more forcible exposition of "Chemico-legal" ink, in its relationship to facts adduced from illustrated scientific testimony, than is to be found in the final opinion written by that eminent jurist Hon. Edgar M. Cullen on behalf of the majority of the Court of Appeals of the State of New York, in the case of De Frees Critten v. The Chemical National Bank. It was the author's privilege to be the expert employed in the lower court about whose testimony Judge Cullen remarks (N. Y. Rep., 171, p. 223) "The alteration of the checks by Davis was established beyond contradiction," and again, p. 227, "The skill of the criminal has kept pace with the advance in honest arts and a forgery may be made so skillfully as to deceive not only the bank but the drawer of the check as to the genuineness of his own signature." The main facts are included in the portion of the opinion cited:

"The plaintiffs kept a large and active account with the defendant, and this action is to recover an alleged balance of a deposit due to them from the bank. The plaintiffs had in their employ a clerk named Davis. It was the duty of Davis to fill up the checks which it might be necessary for the plaintiffs to give in the course of business, to make corresponding entries in the stubs of the check book and present the checks so prepared to Mr. Critten, one of the plaintiffs, for signature, together with the bills in payment of which they were drawn. After signing a check Critten would place it and the bill in an envelope addressed to the proper party, seal the envelope and put it in the mailing drawer. During the period from September, 1897, to October, 1899, in twenty-four separate instances Davis abstracted one of the envelopes from the mailing drawer, opened it, obliterated by acids the name of the payee and the amount specified in the check, then made the check payable to cash and raised its amount, in the majority of cases, by the sum of $100. He would draw the money on the check so altered from the defendant bank, pay the bill for which the check was drawn in cash and appropriate the excess. On one occasion Davis did not collect the altered check from the defendant, but deposited it to his own credit in another bank. When a check was presented to Critten for signature the number of dollars for which it was drawn would be cut in the check by a punching instrument. When Davis altered a check he would punch a new figure in front of those already appearing in the check. The checks so altered by Davis were charged to the account of the plaintiff s, which was balanced every two months and the vouchers returned to them from the bank. To Davis himself the plaintiffs, as a rule, intrusted the verification of the bank balance. This work having in the absence of Davis been committed to another person, the forgeries were discovered and Davis was arrested and punished. It is the amount of these forged checks, over and above the sums for which they were originally drawn, that this action is brought to recover. The defendant pleaded payment and charged negligence on plaintiff's part, both in the manner in which the checks were drawn and in the failure to discover the forgeries when the pass book was balanced and the vouchers surrendered. On the trial the alteration of the checks by Davis was established beyond contradiction and the substantial issue litigated was that of the plaintiff's negligence. The referee rendered a short decision in favor of the plaintiffs in which he states as the ground of his decision that the plaintiffs were not negligent either in signing the checks as drawn by Davis or in failing to discover the forgeries at an earlier date than that at which they were made known to them.

"The relation existing between a bank and a depositor being that of debtor and creditor, the bank can justify a payment on the depositor's account only upon the actual direction of the depositor. 'The question arising on such paper (checks) between drawee and drawer, however, always relate to what the one has authorized the other to do. They are not questions of negligence or of liability to parties upon commercial paper, but are those of authority solely. The question of negligence cannot arise unless the depositor has in drawing his cheek left blanks unfilled, or by some affirmative act of negligence has facilitated the commission of a fraud by those into whose hands the check may come.' (Crawford v. West Side Bank, 100 N. Y. 50.) Therefore, when the fraudulent alteration of the checks was proved, the liability of the bank for their amount was made out and it was incumbent upon the defendant to establish affirmatively negligence on the plaintiff's part to relieve it from the consequences of its fault or misfortune in paying forged orders. Now, while the drawer of a check may be liable where he draws the instrument ill such ill incomplete state as to facilitate or invite fraudulent alterations, it is not the law that he is bound so to prepare the cheek that nobody else call successfully tamper with it. (Societe Generale v. Metropolitan Bank, 27 L. T. [N. S.] 849; Belknap v. National Bank of

North America, 100 Mass. 380) In the present case the fraudulent alteration of the checks was not merely in the perforation of the additional figure, but in the obliteration of the written name of the payee and the substitution therefor of the word 'Cash.' Against this latter change of the instrument the plaintiffs could not have been expected to guard, and without that alteration it would have no way profited the criminal to raise the amount. . . ."

A Pinkerton case of international repute, best known as the "Becker" case, included the successful "raising" of a check by chemical means from $12 to $22,000. The criminal author of this stupendous fraud was Charles Becker, "king of forgers," who as an all round imitator of any writing and manipulator of monetary instruments then stood at the head of his "profession." Arrested and taken to San Francisco he was brought to trial. Two of his "pals" turned state's evidence, and Becker was sentenced to a life term. Through an error on the part of the trial judge he secured a new trial on an appeal to the Supreme Court. The jury disagreed on a second trial, but on the third trial he was convicted. Becker pleaded for mercy, and as he was an old man and showed signs of physical breakdown, the court was lenient with him. Seven years was his sentence.

After his incarceration in San Quetin prison, he described in one sentence how he had risen to the head of the craft of forgers. "A world of patience, a heap of time, and good inks,—that is the secret of my success in the profession."

On completing his sentence, his reply to the question, "What was the underlying motive which induced you to forge?" was one word, "Vanity!"

The detailed facts which follow are from the "American Banker:"

"On December 2, 1895, a smooth-speaking man, under the name of A. H. Dean, hired an office in the Chronicle building at San Francisco, under the guise of a merchant broker, paid a month's rent in advance, and on December 4 he went to the Bank of Nevada and opened an account with $2,500 cash, saying that his account would run from $2,000 to $30,000, and that he would want no accommodation. He manipulated the account so as to invite confidence, and on

December 17 he deposited a check or draft of the Bank of Woodland, Cal., upon its correspondent, the Crocker- Woolworth Bank of San Francisco. The amount was paid to the credit of Dean, the check was sent through the clearing-house, and was paid by the Crocker-Woolworth Bank. The next day, the check having been cleared, Dean called and drew out $20,000, taking the cash in four bags of gold, the teller not having paper money convenient. He had a vehicle at the door, with his office boy inside as driver, and away he went. At the end of the month, when the Crocker-Woolworth Bank made returns to the Woodland Bank, it included the draft for $22,000. Here the fraud was discovered, and here the lesson to bankers of advising drafts received a new illustration. The Bank of Woodland had drawn no such draft, and the only one it had drawn which was not accounted for was one for twelve dollars, issued in favor of A. H. Holmes to an innocent- looking man, who, on December 9, called to ask how he could send twelve dollars to a distant friend, and whether it was better to send a money order or an express order. When he was told he could send it by bank draft, he seemed to have learned something new; supposed that he could not get a bank draft, and he took it, paying the fee. Here came back that innocent twelve-dollar draft, raised to $22,000, and on its way had cost somebody $20,000 in gold.

"The almost absolute perfection with which the draft had been forged had nearly defied the detection of even the microscope. In the body of the original $12 draft had been the words, 'Twelve .... Dollars.' The forger, by the use of some chemical preparation, had erased the final letters 'lve' from the word 'twelve,' and had substituted the letters 'nty-two,' so that in place of the 'twelve,' is it appeared in the genuine draft, there was the word 'twenty-two' in the forged paper.

"In the space between the word 'twenty-two' and the word 'dollars' the forger inserted the word 'thousand,' so that in place of the draft reading 'twelve dollars,' as at first, it read 'twenty-two thousand dollars,' as changed.

"In the original $12 draft, the figures '1' and '2' and the character '$' had been punched so that the combination read '$12.' The forger had filled in these perforations with paper in such away that the part filled in looked exactly like the field of the paper. After having filled in the perforations, he had perforated the paper with the combination, '$22,000.'

"The dates, too, had been erased by the chemical process, and in their stead were dates which would make it appear that the paper bad been presented for payment within a reasonable length of time after it had been issued. The dates in the original draft, if left on the forged draft, would have been liable to arouse suspicion at the bank, for they would have shown that the holder had departed from custom in carrying, such a valuable paper more than a few days.

"That was the extent of the forgeries which had been made in the paper, the manner in which they had been made betrayed the hand of an expert forger. The interjected hand-writing was so nearly like that in the original paper that it took a great while to decide whether or not it was a forgery.

"In the places where letters had been erased by the use of chemicals the coloring of the paper had been restored, so that it was well-nigh impossible to detect a variance of the hue. It was the work of an artist, with pen, ink, chemicals, camel's hair brush, water colors, paper pulp and a perforating machine. Moreover the crime was eighteen days old, and the forger might be in Japan or on his way to Europe. The Protective Committee of the American Bankers' Association held a hurried consultation as soon as the news of the forgery reached New York, and orders were given to get this forger, regardless of expense—he was too dangerous a man to be at large. It was easier said than done; but the skill of the Pinkertons was aroused and the wires were made hot getting an accurate description of Dean from all who had seen him. Suspected bank criminals were shadowed night and day to see if they connected with any one answering the description, but patient, hard labor for nearly two months did not seem to promise much."

Not satisfied with their success in San Francisco these same bank workers began a series of operations in Minneapolis and St. Paul, Minnesota. This information by chance reached the Pinkertons who laid a trap and captured two of the gang. Shortly afterward Becker on information furnished by them was also arrested, taken to California and after three separate trials as before stated, sent to San Quetin.

This triumph of the forger's art, I examined in the city of San Francisco and although it was not, the first time I had been brought into contact with the work of Becker, was compelled to admit that this particular specimen was almost perfect and more nearly so with a single exception than any other which had come under my observation. Becker was a sort of genius in the juggling of bank checks. He knew the values of ink and the correct chemical to affect them. His paper mill was his mouth, in which to manufacture specially prepared pulp to fill in punch holes, which when ironed over, made it most difficult to detect even with a magnifying glass. He was able also to imitate water marks and could reproduce the most intricate designs. He says he has reformed.

During the last twenty years quite a number of cases have been tried in New York City and vicinity in which the question of inks was an all important one. The titles of a few not already referred to are given. herewith: Lawless-Flemming, Albinger Will, Phelan-Press Publishing Co., Ryold, Kerr-Southwick, N. Y. Dredging Co., Thorless-Nernst, Gekouski, Perkins, Bedell forgeries, Storey, Lyddy, Clarke, Woods, Baker, Trefethen, Dupont-Dubos, Schooley, Humphrey, Dietz-Allen, Carter, and Rineard-Bowers.

CHAPTER XXV.

INK UTENSILS OF ANTIQUITY.

THE GRAVING TOOL PRECEDES THE PEN—CLASSIFICATION UNDER TWO HEADS, ONE WHICH SCRATCHED AND THE OTHER WHICH USED AN INK—THE STYLUS AND THE MATERIALS OF WHICH IT WAS COMPOSED—POETICALLY DESCRIBED—COMMENTS BY NOEL HUMPHREYS—RECAPITULATION OF VARIOUS DEVICES BY KNIGHT—BIBLICAL REFERENCES—ENGRAVED STONES AND OTHER MATERIALS THE EARLIEST KINDS OF RECORDS—WHEN THIN BRICKS WERE UTILIZED FOR IN-

SCRIPTION PURPOSES—METHODS EMPLOYED BY THE CHINESE— HILPRECHT'S DISCOVERIES—THE DIAMOND AS A SCRATCHING INSTRUMENT—HISTORICAL INCIDENT WRITTEN WITH ONE—BIBLICAL MENTION ABOUT THE DIAMOND— WHEN IT BECAME POSSIBLE TO INTERPRET CHARACTER VALUES OF ANCIENT HIEROGLYPHICS—DISCOVERY OF THE ROSETTA STONE AND A DESCRIPTION OF IT—SOME OBSERVATIONS ABOUT CHAMPOLLION AND DR. YOUNG WHO DECIPHERED IT—ITS CAPTURE BY THE ENGLISH AND PRESERVATION IN THE BRITISH MUSEUM—EMPLOYMENT OF THE REED PEN AND PENCIL- BRUSH—THE BRUSH PRECEDED THE REED PEN—THE PLACES WHERE THE REEDS GREW—COMMENTS BY VARIOUS WRITERS—METHOD OF FORMING THE REED INTO A PEN—CONTINUED EMPLOYMENT OF THEM IN THE FAR EAST—THE BRUSH STILL IN USE IN CHINA AND JAPAN—EARLIEST EXAMPLES OF REED PEN WRITING—WHEN THE QUILL WAS SUBSTITUTED FOR THE REED—REED PENS FOUND IN THE RUINS OF HERCULANEUM—ANECDOTE BY THE ABBE, HUC.

THE instruments of antiquity employed in the art of writing belong to two of the most distant epochs.

In the first period, inscriptions were engraved, carved or impressed with sharp instruments, and of patterns characteristic of a graving tool, chisel or other form which could be adapted to particular substances like stone, leaves, metal or ivory plates, wax or clay tablets, cylinders and prisms.

The ancient Assyrians even used knives or stamps for impressing their cuneiform writing upon cylinders or prisms of soft clay which were often glazed by subsequent bakings in kilns.

The other period was that in which written characters were made with liquids or paints of any kind or color. The liquids (inks) were used in connection with a pen manufactured from a reed (calamus), while the paints were "painted" on the various substances with a brush. The writing executed with both of these instruments was on materials like the bark of trees, cloth, skins, papyrus, vellum, etc.

The ancient as well as modern pens, though of many sorts and kinds, are to be classified under two general heads, those which scratch and those which use an ink.

There is no authority to dispute the generally conceded fact that the "scratching" instrument was the first one used. Its most popular form seems to have been the stylus or bodkin, which was made of a variety of materials, such as iron, ivory, bone, minerals or any other hard substance, which could be sufficiently sharpened at one end to indent the various materials employed in connection with its use. The other end was flattened for erasing marks made on wax and smoothing it. From it the Italian stilletto took its origin.

The stylus is best described in the following lines:

"My head is flat and smooth, but sharp my foot,
And by man's hand to different uses put;
For what my foot performs with art and care,
My head makes void, such opposites they are."

Relative to the employment of marking instruments which belong to the most venerable antiquity, Noel Humphreys observes:

"Before the growth of wealth and luxury had taught nations to raise magnificent temples and stately palaces, whose walls the hieroglyphic sculptor covered with records of the pomp and pride of princes, more purely national memorials had found their place upon the native rock, the most convenient surfaces of which were smoothed for this purpose. Where no such rock existed in the situation required, a massive stone was raised by artificial means and the record, whether referring to a victory, a new boundary, or any other event of national interest was engraved upon it. Such memorials have been described by Hebrew writers as aumad or ammod, literally, the lips of the people, or, the words of the people, but actually meaning a pillar. Records in this form and the early name they bore account for the strange legends of mediaeval times referring to speaking stones—a name by which such monuments were probably still called long after time had effaced the speaking record, and the original purport of the defaced stone was forgotten. In semi-barbarous epochs, like the era which followed the partial extinction of Roman civilization, popular curiosity and superstition combined would seek to give a meaning to the name of such 'speaking stones,' and as an example of the legends which thus arose, the itinerarium cambriae of Geraldus may be cited, in which a stone is mentioned at St. David's as the 'speaking stone' (lech lavar) which was said to call out when a dead body was placed upon it. The most remarkable rock inscriptions still remaining are those of Assyria and Persia, but many national tablets of more recent date are still in existence. For the execution of such records and those of the palaces of Egypt and Assyria, some kind of steel point must have been used, as no softer substance would have served to engrave them in granitic and basaltic slabs with the sharpness they still exhibit, which proves that the art of hardening steel, long thought a comparatively modern invention, was known to the ancient people of Asia and Africa."

A list of the various devices of different countries, by which characters could be legibly portrayed with a scratching implement, is best recapitulated by Mr. Knight, who presents them in the following order:

"The tabula or wooden board smeared with wax, upon which a letter was written by a stylus.

"The Athenian scratched his vote upon a shell as did the lout when he voted to ostracize Aristides.

"The records of Ninevah were inscribed upon tablets of clay, which were then baked.

"The laws of Rome were engraved on brass and laid up in the Capitol.

"The decalogue was graven upon the tables of stone.

"The Egyptians used papyrus and granite.

"The Burmese, tablets of ivory and leaves.

"Pliny mentions sheets of lead, books of linen, and waxed tablets of wood.

"The Hebrews used linen and skins.

"The Persians, Mexicans, and North American
Indians used skins.

"The Greeks, prepared skins called membrana.

"The people of Pergamus, parchment and vellum.

"The Hindoos, palm-leaves."

The written deeds of biblical time were kept in various styles of pottery (Jeremiah xxxii. 14). Handwriting on tiles was common in Egypt, Assyria and Palestine (Ezekiel iv. I). Such handwritings were on tablets of terra-cotta or common baked clay bricks. One of the kind was fashioned by inscribing directly with a "stylus" on the clay, before baking. Another, were "moulds" made from older inscriptions or duplicates from the first kind.

The Hebrew term sepher, translated into English means a "book," and some authorities claim it is derived from the same root as the Greek <gr kefas>, a stone, which would seem to point to engraved stones as the earliest kinds of records. Indeed nearly all the passages in the Five Books of Moses, in which writing is mentioned, refer to records of this kind, or to tablets of lead or wood, occasionally described as coated with wax.

Long before the use of papyrus, or any like substance was known as a material for writing on, thin bricks were frequently utilized for such purposes. The Chinese wrote on slips of bamboo which had been previously scraped to be afterwards submitted to intense heat which so hardened them, that a graver would cut lines with the same facility, as could be accomplished on soft metal like lead. These bamboo tablets were joined together by means of cords made of bark and when folded formed a "book." Different nations adopted other modes in their preparation of surfaces to engrave on. Many original specimens have come down to us which present definite evidence of the variety of materials and methods employed in their manufacture.

Hilprecht, "Explorations in Bible Lands," 1903, mentions many discoveries of such specimens. He says that more than four thousand clay tablets were discovered during the excavations of 1889 and 1900.

These relics call attention only to a very few discoveries of this character. There were other explorers who preceded Hilprecht in this direction, and who with him have thus secured tangible evidence which fully confirms all that has been said about the employment of the most ancient of writing instruments, the "stylus."

The diamond is also to be classified under the head of "scratching implements" and many historical incidents are recorded of its use. One of the most interesting relates to Sir Walter Raleigh and Queen Elizabeth and to be found in Scott's "Kenilworth." Sir Walter, using his diamond ring, wrote on a pane of glass in her summer-house at Greenwich:

"Fain would I climb, but that I fear to fall."

The maiden Queen adding the words:

"If thy mind fail thee, do not climb at all."

Biblical mention of the diamond, employed as a pen, is found in Jeremiah xvii. 1.

> "The sin of Judah is written with a pen of iron,
> and with the point of a diamond."

It has not always been possible to decipher and interpret the character values of the most ancient hieroglyphics or picture writings inscribed on bricks, stone and metal slabs, and the Egyptian monuments. The means to do so were furnished as the result of a very fortunate accident or "find."

A French artillery officer in 1799 while excavating the foundations for a fortification near the Rosetta mouth of the Nile, found a curious black tablet of stone. On it were engraved three inscriptions, each of different characters and dialects.

The first of the three inscriptions was in hieroglyphic, then unreadable; the second in demotic or shorter script, also unknown, and the third in a living language pertaining to the time of Ptolemy Epiphanes, who reigned about 200 B. C.

This relic of antiquity is called the Rosetta stone.

Jean Francois Champollion, who with Dr. Thomas Young studied the intricacies of these writings, first established the fact that the three inscriptions on this stone were translations of each other. Dr. Young's investigations caused him to study the language included in the second inscription, and made his deductions, it is said, "by dint of thousands of scientific guesses, all but a few of which were eliminated by tests which he invented and applied; he at last discovered and put together the set of fundamental principles that govern the ancient writings."

Champollion, however, began at the bottom and having successfully translated the LIVING language, established a "key" or alphabet. Hence it became possible, although requiring some years, to solve the mystery of writings of 4000 or more years old.

Champollion pursued his discoveries so thoroughly in this direction as to be able to complete in 1829 an Egyptian vocabulary and grammar.

The Rosetta stone after remaining in the possession of the French for many years was captured by the English on the defeat of the French forces in Egypt and is now in the British museum.

As writing with liquid colors on papyrus or analogous materials which could be used in the form of rolls, gradually came into vogue, the calamus or reed pen, pencil brush (hair pencil), or the juncas, a pen formed from a kind of cane, were more or less employed.

The "calamus" followed the "brush," just as

phonographic writing which denotes arbitrary sounds or the language of symbols, came after the picture or ideographic writing.

The places where the calamus grew and the modes of preparing them are variously discussed by different ancient and modern writers. Some claim that the best reeds for pen purposes formerly grew near Memphis on the Nile, near Cnidus of Caria, in Asia Minor, and in Armenia. Those grown in Italy were estimated to have been of but poor quality. Chardin calls attention to a kind to be found, "in a large fen or tract of soggy land supplied with water by the river Helle, a place in Arabia formed by the united arms of the Euphrates and Tigris. They are cut in March, tied in bundles, laid six months in a manure heap, where they assume a beautiful color, mottled yellow and black." Tournefort saw them growing in the neighborhood of Teflis in Georgia. Miller describes the cane as "growing no higher than a man, the stem three or four lines in thickness and solid from one knot to another, excepting the central white pith."

The incipient fermentation in the manure heap dries up the pith and hardens the cane. The pens were about the size of the largest swan's quills. They were cut and slit like a quill pen but with much larger nibs.

In the far East the calamus is still used, the best being gathered in the month of March, near Aurac, on the Persian Gulf, and still prepared after the old method of immersing them for about six months in fermenting manure which coats them with a sort of dark varnish and the darker their color the more they are prized.

The "brush" also holds its career of usefulness, more especially in China and Japan.

The earliest examples of reed pen writing are the ancient rolls of papyrus which have been found buried with the Egyptian dead. Some of these old relics of antiquity are claimed to have been prepared fully twenty centuries or more before the Christian era.

The "reed" pen for ink writing held almost undisputed sway until the sixth century after the Christian era, when the quill (penna) came into vogue.

Reed pens preserved in excellent condition were found in the ruins of Herculaneum.

"When he had finished, he dried the bamboo-pen on his hair, and replaced it behind his ear, saying, 'Yak pose' (That is well). 'Temou chu' (Rest in peace), we replied; and, after politely putting out our tongues, withdrew." Abbe Hue at Lha-Ssa.

CHAPTER XXVI.

INK UTENSILS (QUILL PEN STEEL PEN).

THE QUILL PEN THE MOST SUCCESSFUL AND FITTING OF ALL WRITING INSTRUMENTS—TENDENCY TO "WEAR" OUT—THE SOMETIMES AFFECTION FOR OLD PENS—DR. HOLLAND'S LINES ON THE PEN—SELECTION OF QUILLS TO BE MADE INTO PENS—METHOD OF PREPARING

THEM—BYRON'S ESTIMATION OF HIS QUILL PEN—ITS INVENTION BEFORE THE SIXTH CENTURY UNCERTAIN—EMPLOYMENT OF THE REED AND QUILL PEN TOGETHER UNTIL THE TWELFTH CENTURY—WHEN THE STEEL PEN CAME INTO VOGUE—WHO WAS ITS INVENTOR—SOME OBSERVATIONS ABOUT IT—QUANTITY OF MATERIAL SIXTY YEARS AGO CONSUMED IN PEN MANUFACTURE—A FEW REMARKS ABOUT GOLD, FOUNTAIN AND STYLOGRAPHIC PENS—MORE STEEL USED IN THE MANUFACTURE OF PENS THAN IN THAT OF SWORDS AND GUNS—POETICAL LINES ABOUT THE PEN.

THE quills belonging to the feathers of birds seem to have been the most successful and fitting of all materials for pens, for, though steel and other metals are now used for this purpose to an immense extent, there is a power of adaptation in a quill pen which has never yet been equalled in metal. Quills, however, like other things, have a tendency to "wear out," and the trouble resulting from the necessity of frequently mending quill pens and a desire to write with more rapidity have been the main causes of the introduction of steel substitutes. A kind of affection has often been felt by an author or official, or their admirers, for the pen with which he has written any large or celebrated work or signed some important document; old worn-out pens, as well as new ones, have been preserved as memorials in connection with such matters, and Dr. Holland, who translated Pliny's "Natural History" in the sixteenth century, recorded an exploit connected with it in the following lines:

"With one sole pen I wrote this book,
Made of a gray goose-quill:
A pen it was when it I took
A pen I leave it still."

The quills employed for pens were generally those of the goose, although the crow, the swan, and other birds yielded feathers which were occasionally available for this purpose. Each wing produced about five good quills, but the number thus yielded was so small that the geese reared in England could not furnish nearly enough for the demand, hence the importation of goose quills from the Continent was very large. The process surrounding the manufacture of a quill pen proves of considerable interest.

"The geese are plucked of their feathers three or four times a year, the first time for the sake both of the quills and the feathers, but the other times for the feathers only. The pen quills are generally taken from the ends of the wings. When plucked the quills are found to be covered with a membranous skin, resulting from a decay of a kind of sheath which had enveloped them; the interior vascular membrane, too, resulting from the decay of the vascular pith, adheres so strongly to the barrel of the quill as to be with difficulty separated, while, at the same time, the barrel itself is opaque, soft, and tough. To remove these various defects the quills undergo several processes. In the first instance, as a means of removing the membraneous skin, the quills are plunged into heated sand, the high temperature of which causes the external skin of the barrel to crack and peel off, and the internal membrane to shrivel up. The outer membrane is then scraped off with a sharp instrument, while the inner membrane remains in a state to be easily detached. For the finest quills the heating is repeated two or three times. The heat of the sand, by consuming or drying up the natural moisture of the barrel, renders it harder and more transparent. In order to give the barrel a yellow color, and a tendency to split more readily and clearly, it is dipped in weak nitric acid, but this was considered to render the quill more brittle and less durable, and was therefore a sacrifice of utility for the sake of appearance."

"Oh! nature's noblest gift—my gray goose quill!
Slave of my thoughts, obedient to my will,
Torn from thy parent bird to form a pen,
That mighty instrument of little men!"
BYRON.

To locate an exact period for the invention of the quill pen is impossible. It could hardly have been in use before the fourth century, probably not earlier than two centuries later. Some writers have assumed that it was employed by the Romans, but as no distinct mention is made of them by early classical authors we must accept the only information at hand.

Isidore (died A. D. 636) and contemporaries state that the quills of birds came into use as pens only in the sixth century. It is also known, St. Brovverus being the authority, that in his time (seventh century) the calamus or reed pen and the quill pen were employed together, the calamus being used in the writing of the uncial (inch) letters and capitals, and the quill for smaller letters. Mention is also made by many writers of the five centuries which followed Isidore's time of the calamus, indicating that notwithstanding it had been superseded by the quill it was still a favorite writing implement in some places.

The use of the "steel pen" did not spring immediately from that of the "quill pen." There were several intermediate stages adopted before the fitness of steel for this purpose was sufficiently known, From about 1800 to 1835 the number of proposed substitutes for the quill pen was very considerable. Horn pens, tortoise-shell pens, nibs of diamond or ruby imbedded in tortoise shell, nibs of ruby set in fine gold, nibs of rhodium and of iridium imbedded in gold,— all have been adopted at different times, but most of them have been found too costly for general adoption. Steel is proved to be sufficiently elastic and durable to form very good pens, and the ingenuity of manufacturers has been exerted to give to such pens as many as possible of the good qualities possessed by the quill pen.

The original flexible iron pen of modern times was an experimental affair probably, being mentioned by Chamberlayne as far back as 1685.

The first steel pens in regular use were made by
Wise, in London, in 1803, and for many years thereafter.

His pen was made with a barrel, by which it was slipped upon a straight handle. In its portable form it was mounted in a bone case for the pocket. Prejudice, however, was strong against them, and up to 1835 or thereabouts quills maintained their full sway, and much later among the old-fashioned folks. To him, however, is due the credit of being the inventor of the modern steel pen.

It has been the thought of some people that Gillott was the progenitor of the steel pen, but he was not. Arnoux, a French mechanic, made metallic pens with side slits in 1750. Samuel Harrison, an

Englishman, made a steel pen for Dr. Priestly in 1780. Peregrine Williamson, a native of New York, while engaged as a jeweler in the city of Baltimore, made steel pens in 1800.

Perry's first pens were of steel, rolled from wire, the material costing seven shillings a pound. Five shillings each was paid the workman for making them; this was afterward reduced to thirty-six shillings per gross, which price was continued for several years.

It was Joseph Gillott, however, originally a Sheffield cutler, and afterwards a workman in light steel articles, as buckles, chains, and other articles of that class, who in 1822 gave impulse to the steel-pen manufacture. Previous to his entering the business the pens were cut out with shears and finished with the file. Gillott adapted the stamping press to the requirements of the manufacture, as cutting out the blanks, forming the slits, bending the metal, and impressing the maker's name on the pens. He also devised improved modes of preparing the metal for the action of the press, tempering, cleansing, and polishing, and, in short, many little details of manufacture necessary to give them the required flexibility to enable them to compete with the quill pen. One great difficulty to be overcome was their extreme hardness and stiffness; this was effected by making slits at the side in addition to the central one, which had previously been solely used. A further improvement, that of cross grinding the points, was subsequently adopted. The first gross of pens with three slits was sold for seven pounds. In 1830 the price was $2.00; in 1832, $1.50; in 1861, 12 cents, and a common variety for 4 cents a gross. About 9,300 tons of steel are annually consumed, the number of pens produced in England alone being about 8,000,000,000.

Bramah patented quill nibs made by splitting quills and cutting the semicylinders into sections which were shaped into pens and adapted to be placed in a holder. These were, perhaps, the first nibs, the progenitors of a host of steel, gold, and other pens.

Hawkins and Mordan, in 1823, made nibs of horn and tortoise shell, instead of quill. The tortoise shell being softened, points of ruby and diamond were imbedded. Metallic points were also cemented to the shell nibs.

Doughty, about 1825, made gold pens with ruby points.

Gold pens with rhodium or iridium points were introduced soon afterwards.

Mordan's oblique pen, English patent, 1831, was designed to present the nibs in the right direction while preserving the customary positions of the pen and hand.

The fountain pen carries a supply of ink, fed gradually to the point of the instrument. The first made by Scheffer was introduced about 1835 by Mordan. The pressure of the thumb on a stud in a holder caused a continuous supply of ink to flow from the reservoir to the pen.

The "stylographic" is a reservoir pen shaped like a pencil, in which the flow of ink is regulated by pressure of a style or fine needle with blunt point upon the paper. It must be held in a vertical position. All marks made with one, both up and down strokes, are equal in width.

Gold pens are now usually tipped with iridium, making what are commonly known as diamond points.

"The iridium for this purpose is found in small grains of platinum, slightly alloyed with this latter metal. The gold for pens is alloyed with silver to about sixteen carats fineness, rolled into thin strips, from which the blanks are struck. The under side of the point is notched by a small circular saw to receive the iridium point, which is selected with the aid of a microscope. A flux of borax and a blowpipe secure it to its place. The point is then ground on a copper wheel of emery. The pen-blank is next rolled to the requisite thinness by the means of rollers especially adapted for the purpose, and tempered by blows from a hammer. It is then trimmed around the edges, stamped, and formed in a press. The slit is next cut through the solid iridium point by means of a thin copper wheel fed with fine emery, and a saw extends the aperture along the pen itself. The inside edges of the slit are smoothed and polished by the emery wheel; burnishing and hammering produce the proper degree of elasticity."

It is asserted that more steel is used in the manufacture of pens than in all the swords and guns in the world. This fact partly verifies the old saying, "The pen is mightier than the sword."

"Three things bear mighty sway with men,
The Sword, the Sceptre, and the Pen;
Who can the least of these command,
In the first rank of Fame will stand."

CHAPTER XXVII.

SUBSTITUTES FOR INK UTENSILS ("LEAD" AND OTHER PENCILS).

"BLACK-LEAD" PENCILS AN EXCELLENT PEN SUBSTITUTE UNDER CERTAIN CONDITIONS—ITS COMPOSITION—"BLACK-LEAD" CONTAINS NO LEAD, HENCE THE NAME IS MISAPPLIED—THE DISCOVERY OF ITS PRINCIPAL SOURCE OF SUPPLY AN ACCIDENT—A DESCRIPTION OF HOW IT IS MINED—TREATMENT BEFORE BEING INTRODUCED INTO THE GROOVED WOOD—USE OF RED AND BLACK CHALK PENCILS IN GERMANY, 1450—THEIR USE IN MEXICO IN EARLY TIMES—WHO MANUFACTURES LEAD PENCILS—EMPLOYMENT OF THE COMPOSITION OF LEAD AND TIN IN MEDIAEVAL TIMES—BAVARIAN GOVERNMENT IN 1816 A MANUFACTURER OF LEAD PENCILS.

THE black-lead pencil, under many circumstances, is a very useful substitute for the pen, in that it requires no liquid ink for marking the characters on paper or other materials. The peculiar substance which fills the central channel of the stick of cedar has the property of marking when it touches paper; and, as the marks thus made are susceptible to easy removal, a pencil of this kind is available for purposes which would not be answered by the use of pen and ink.

The substance misnamed "black-lead" contains NO LEAD and is a carburet of iron, being composed of carbon and iron. It generally occurs in Mountain districts, in small kidney-shaped pieces, varying in size from that of a pea upwards, which are interspersed among various strata, and is met with in different parts of the world.

Its principal source of supply until about 1845, when it became exhausted, was the Borrowdale mine in Cumberland, England, which was discovered in 1564. About 1852 a number of mines were opened containing this substance in Siberia and from which place the best products are now obtained.

The accidental discovery of this mineral at Borrowdale was during the reign of Queen Elizabeth who made many inquiries about it. The name of this mineral was locally known as wad (graphite). So valuable was it regarded that it commanded a very high price, and this price acted as in inducement to the workmen and others to pilfer pieces from the mine. For a number of years scenes of great commotion took place, arising out of these depredations; and the result was that the proprietors adopted such stringent rules that hardly anything was known of the internal economy of the mine till about sixty years ago, when Mr. Parkes gave a description of it, from which I may condense a few particulars.

The mine is in the midst of a mountain about two thousand feet high, which rises at in angle of about 45 degrees; and, as that part of the mine which has been worked during the last century is near the middle of the mountain, the present entrance is about a thousand feet from the summit. The opening by which the workmen enter descends by a flight of steps; and in order to guard the treasure within, the proprietors have erected a strong brick building of four rooms, one of which is immediately over the entrance into the mine. This entrance is secured by a trap-door, and the room connected with it serves as a dressing-room for the men when they enter and leave the mine. The men work in gangs, which relieve each other every six hours, and when the hour of relief comes, a steward or foreman attends the dressing- room to see the men change their dresses as they come up one by one out of the mine. The clothes are examined by the steward to see that no black-lead is concealed in them; and when the men have dressed they leave the mine, making room for another gang, who change their clothes, enter the mine, and are fastened in for six hours. In one of the four rooms of which the house consists there is a table, at which men are employed in sorting and dressing the mineral. This is necessary, because it is usually divided into two qualities, the finest of which have generally pieces of iron- ore or other impurity attached to them, which must be dressed off. These men, who are strictly watched while at work, put the dressed black-lead into casks holding about one hundred-weight each, in which state it leaves the mine. The casks are conveyed down the side of the mountain in a curious manner. Each cask is fixed upon a light sledge with two wheels, and a man, who is well used to the precipitous path, walks down in front of the sledge, taking care that it does not acquire momentum enough to overpower him. When the cask has been thus guided safely to the bottom, the man carries the sledge up hill upon his shoulders, and prepares for another descent.

Up to about the middle of the eighteenth century the mine was opened only once in seven years, the quantity taken out at each time of opening being such as was deemed sufficient to serve the market for seven years; but when, at a later period, it was found that the demand was increasing and the supply decreasing, it was deemed necessary to work the mine six or seven weeks every year. During the time of working, the mine is guarded night and day; and when a quantity sufficient for one year's consumption has been taken out, the mine is secured until the following year. Several hundred cart-loads of rubbish are wheeled into the mine, so as to block up the entrance completely; and this rubbish acts as a dam to prevent the springs and land waters from flowing out, so that the mine gradually becomes flooded.

When the Year's mining is concluded, the barrels of black-lead are brought to market, and the mode of effecting the sales was described by Dr. Faraday some years ago to be as follows: A market is held on the first Monday of every month at a house in London, where the buyers, who are generally only seven or eight in number, examine each piece with a sharp instrument to ascertain its hardness, those which are too soft being rejected. The person who has the first choice pays 45s. per pound, the others 30s. But, as there is no addition made to the first quantity in the market, the residual portions are examined over and over again until they are exhausted. At one time the annual sale was said to amount to the value of L40,000 per annum, but it has been greatly reduced since.

A mode of applying manufacturing processes to the preparation of black-lead is described by Dr. Ure as being adopted in Paris. The mineral, being reduced to a fine powder, is mixed with very pure powdered clay, and the two are calcined in a crucible at a white heat; the proportion of clay employed is greater as the pencil is required to be harder, the average being equal parts of both. The ingredients are ground with a muller on a porphyry slab and then made into balls, which are preserved in a moist atmosphere in the form of paste. The paste is pressed into grooves cut in a smooth board, and another board, previously greased, is pressed down upon it. When the paste has had time to dry, the mould or grooved board is put into a moderately heated oven, by which the paste, now in the form of square pencils, shrinks sufficiently to fall out of the grooves. In order to give solidity to the pencils they are set upright in a crucible and surrounded with pounded charcoal, fine sand, or sifted ashes; the crucible, being covered, is exposed to a degree of heat proportionate to the hardness required in the pencils, the harder pencils requiring the higher degree of heat. Some of the pencils are shaped in a curious manner: models of the pencils, made of iron, are stuck upright upon an iron tray, having edges raised as high as the intended length of the pencils; and a metallic alloy, made of tin, lead, antimony and bismuth is poured into the sheet-iron tray. When the alloy has cooled, it is inverted and shaken off from the model-rods, so as to form a mass of metal perforated throughout with tubular cavities corresponding in size with the intended pencil pieces; the pencil paste is introduced by pressure into these cavities, and when nearly dry the pieces shrink sufficiently to be easily removed from the cavities.

The pencils just described are alike throughout all their thickness, but in the majority of English pencils there is a wooden holder to contain a narrow filament of black lead running down the middle. So long ago as the year 1618 this mode was adopted; for Sir John Pettus, who was deputy governor of the Borrowdale mine under Charles II, in his "Fleta Minor," while, speaking of black-lead says, that "Of late it is curiously formed into cases of deal or cedar and so sold as dry pencils, something more useful than pen and ink." In a general way modern black-lead pencils, are made by sawing cedar first into long planks, and then into smaller rods; grooves are cut out by means of a cutting machine moved by a fly- wheel to such a depth as will receive a small layer of black-lead; the pieces of the mineral are cut into thin slabs and then into rods the same size as the grooves, into which they are inserted; the two halves of the case are then glued together, and the whole is turned into a cylindrical form by means of a guage.

The kind of pencil called "crayon" is a mixture of some kind of earth with a coloring substance. The earth employed is sometimes chalk, and at other times pipe-clay, gypsum, starch-flour, or ochre. The coloring substance is yellow ochre, mineral yellow, chrome, red chalk, vermilion, indigo—indeed, any of the usual dry colors, according to the tint required. Besides the earth and the color, there is a gummy liquid required to combine them together; gum arabic, gum tragacanth, and in some cases oil, wax, or suet, are used as the third ingredient. The crayons here alluded to are employed rather for drawing than for writing, but they obviously belong to the class of pencils in their mode of action.

The ancients drew lines and letters with wooden styles, and afterward an alloy of lead and tin was used. Pliny refers to the use of lead for ruling lines on papyrus. La Moine cites a document of 1387 ruled with graphite. Slips of graphite in wooden sticks (pencils) are mentioned by Gesner, of Zurich, in 1565; he credits England with the production. They are doubtless the product of the Borrowdale mine, then lately discovered. In the early part of the seventeenth century black-lead pencils are distinctly described by several writers. They are noticed by Ambrosinus, 1648; spoken of by Pettus, in 1683, as inclosed in fir or cedar.

Red and black chalk pencils were used in Germany in 1450; in fact, fragments of chalk, charcoal, and shaped sticks of colored minerals had been in use since times previous to all historic mention.

When Cortez landed in Mexico, in 1520, he found the Aztecs using graphite crayons, which were probably made from a mineral found in Sonora.

The firm of A. W. Faber are the largest manufacturers of lead pencils in the world. They have compiled a history of this implement of handwriting which they have permitted me to use in the story which follows.

The lead pencil is an invention of modern times, and its introduction may deservedly be ranked with the large number of technical innovations in which more especially the last three centuries have been so rich; nor can it be denied that pencils have played an important part in the diffusion of arts and sciences and in facilitating study and intellectual intercourse.

To the classic ages and their art the pencil, and in general every application of lead as a writing material, was entirely unknown, and it was not till the advent of the middle ages that it began to be used for this purpose. This lead, i. e. metallic lead, however, was in no way equivalent to the graphite or black-lead of our pencils, which are only honored with the prefix of "lead," owing to the leaden color of the writing done with them.

Moreover, in those days, lead was used exclusively for ruling and in no way for writing or drawing; it was employed in the form of round, sharp-edged discs, similar to those which, it is said, were already used for the same purpose in ancient classic times. It is only with the development and growth of modern painting that traces of pencil-like drawings first begin to be met. At so early a period even as the fourteenth century, mention is made by the masters of that time, more especially by the brothers Van Eyck, and again in the fifteenth century by Menlink and others, of studies or compositions which were made with an instrument similar to a lead pencil, upon a paper with chalk prepared surface.

This type of drawing was commonly classed as "silver- style," a term, however, which was no doubt erroneous, as there could be no question of the use of pure silver in this connection.

In the same way it is also reported of the later mediaeval Italian artists that they drew their subjects in "silver-style," upon planished fig-tree wood, the surface of which had been prepared with the powder obtained from calcined bones,—a method, however, which seems only to have been employed in exceptional instances.

But in the fourteenth century, drawings were frequently done in Italy with pencils consisting of a mixture cast from lead and tin; these drawings could easily be erased with bread crumbs.

Petrarch's "Laura" was portrayed in this manner by one of his contemporaries, and the method was still in vogue in the days of Michael Angelo. From Italy these pencils subsequently found their way to Germany, but it is not apparent under what particular name. In Italy itself they were called "stili," the equivalent of the word stylus. At no time, however, do these varieties seem to have been the predominating material used for drawing purposes.

In conjunction with these, pens were used for writing and drawing, and at the zenith of the art period of those days black and red crayons were also used on a large scale. The Italians imported the best qualities of red crayons from Germany, the best black chalk being obtained from Spain.

Vasari writes of a certain sixteenth century artist, that he was equally skillful in handling the stylus or the pen, black chalk or red crayon.

It was this period which witnessed the discovery of plumbago, a mineral which was soon worked up into an entirely new material for writing and drawing,— the lead pencil.

This discovery, which was destined to confer such great benefits not only upon practical life, but also upon art, was made in England during the reign of Queen Elizabeth, for in the year 1564 the celebrated black-lead mines of Borrowdale, in Cumberland, were discovered. With the opening of this mine, the first material steps were taken to implant on English soil a lead pencil industry which in the course of time was to assume important dimensions.

The first lead pencils are supposed to have been manufactured in England in the second half of the sixteenth century. The raw plumbago, or "wad," as it was locally termed, was subjected to the following treatment: "On reaching the surface it was sawn into strips of the required size, and these, without any further manipulation, were inserted into the wood. Strange though it may appear, the lead pencils first manufactured in this manner are acknowledged to have been the best—and even at the beginning of the present century they remained unsurpassed upon the score of the softness and fine tone of the lead. Although the Cumberland lead pencils were in great demand owing to the fact that they were the first to successfully meet a long-felt want, they nevertheless owed their permanent and wide-spread reputation— more especially in artistic circles—to their excellent quality.

Towards the end of the last century the black-lead pencil industry was introduced into France, where with some restrictions it soon developed.

With the removal of all restrictions on industrial freedom in 1795, the idea was entertained of using clay as a binding medium for black-lead. This method offered several advantages, for not only did the addition of clay cause a saving of a large percentage of the valuable mineral, but it greatly facilitated the method of manufacture, so that lead pencils could now be offered at greatly reduced prices.

By these improvements a new era in the manufacture of lead pencils was begun in France. Still, there remained much to be done in the field of black- lead pencil making in order to do justice to the increasing demands of art and the requirements of more civilized life.

It is true, different kinds of lead pencils of various degrees were produced, but they did not comply by a long way with the different uses for which they were needed. The manipulation of the brittle material required not only deep study, but also conscientious and skillful workmen, in order to impart the necessary standard of perfection to the lead pencil.

Among the various German industries the manufacture of black-lead pencils occupied but a very modest place.

The first traces of its existence are to be found at Stein, a village not far from Nuremberg. As far back as the year 1726 the church registers mention marriages between "black-lead pencil makers," and, at a later date references are found in the same registers to "black-lead cutters" of both sexes.

The manufacture of black-lead pencils, however, occupied a position on the very lowest rung of the industrial ladder.

But is time proceeded the Bavarian government directed their attention to this branch of industry, and did all in their power to encourage it; and, as early as the year 1766, a Count von Kronsfeld obtained a concession to establish a lead pencil factory at Jettenbach.

Later on, in the year 1816, the Bavarian government established a royal lead pencil manufactory at Obernzell (Hafnerzell), and introduced into it the French process, described above, of using clay as a binding medium for graphite.

CHAPTER XXVIII.

ANCIENT INK BACKGROUNDS (THE ORIGIN OF PAPYRUS).

FROM WHENCE COMES THE NAME PAPER—FIRST CENTURY COMMENT ABOUT IT—KNIGHT'S COMMENTS MORE THAN 1,800 YEARS LATER—PAPYRUS AN EGYPTIAN REED—NAMES BESTOWED BY ANCIENT WRITERS—THE SAME NAMES AS EMPLOYED IN MODERN TIMES—LEAVES OF PLANTS PRECEDED THE INVENTION OF PAPYRUS—WHEN IT WAS THAT ROLLED RECORDS CAME INTO VOGUE—VARRO'S ESTIMATION AS TO THE ORIGINAL USE OF PAPYRUS NOT CORRECT—REAL FACTS RESPECTING THE INTRODUCTION OF PAPYRUS BEYOND THE LIMITS OF EGYPT—CHARACTER OF MATERIALS EMPLOYED BY THE GREEKS BEFORE THAT EPOCH—EMPLOYMENT OF IT FOR LITERARY PURPOSES—ADOPTION OF PARCHMENT AND VELLUM—PAPYRUS MSS. EMPLOYED IN THE FORM OF ROLLS AND THE REASON FOR SAME—ANCIENT MANUFACTURE OF PAPYRUS IN EGYPT—SOME OF THE NAMES
USED TO DESIGNATE DIFFERENT KINDS—PLINY'S DESCRIPTION OF THE MANUFACTURE OF PAPYRUS AND HIS MISINFORMATION ABOUT IT—WHERE IT FLOURISHED BEST—PAPYRUS AS KNOWN TO THE HEBREWS AND ITS BIBLICAL MENTION—MANUFACTURE OF PAPYRUS IN THE ANCIENT CITY OF MEMPHIS—CHARACTERISTICS OF THE PAPER EMPLOYED BY THE MEXICANS—MR. HARRIS'S DISCOVERY OF ANCIENT FRAGMENTS OF PAPYRUS—THE STORY ABOUT IT AS TOLD BY THE LONDON ATHENaeUM—DATES OF THE OLDEST KNOWN SPECIMENS OF GREEK PAPYRI—DATE OF THE FIRST DISCOVERY

OF GREEK PAPYRI—USE OF OTHER PLIABLE MATERIALS WITH PAPYRUS—HOW THEY WERE PREPARED FOR WRITING PURPOSES—DOUBTS AS TO TIME THAT ROLLED RECORDS SUPERSEDED TABLET FORMS—SUGGESTIONS BY NOEL HUMPHREYS—VIEWS ENTERTAINED BY EARLIER WRITERS.

THE name paper is derived from papyrus, a reed grown in Egypt, whose stalk furnished for so many centuries the principal material for writing upon to the people of that country and those bordering on the Mediterranean Sea. In the first century of the Christian era the younger Pliny remarks:

"All the usages of civilized life depend in a remarkable degree upon the employment of paper. At all events, the remembrance of past events."

A statement which has caused Mr. Knight to make the following comment:

"This observation, undoubtedly true 1,800 years ago, is much more remarkably so now; indeed, in considering that paper as we now understand it was entirely unknown to Europe in the time of Pliny, the expression of the great dependence upon what seems to us so fragile and inefficient a substitute for real paper appears strange."

Mr. Knight also says that the Greek name papuros, mentioned by Theophrastus, a contemporary of Aristotle and Alexander, was probably the Egyptian name of the reed with a Greek termination. It was also called biblos by Homer and Herodotus, whence our term bible. The term volumen, a scroll, indicates the early form of a book of bark, papyrus, skin, or parchment, as the term liber (Latin, a book, or the inner bark of a tree) does the use of the bark itself. Hence also our terms library and librarian. "Book" is also derived from the Danish word bog, the bark of the beech. Pliny quoting Varro, who preceded him some two centuries, asserts that before the invention of papyrus, the large leaves of certain plants were prepared so that they could be written upon. Hence originates our term "leaves" of a book which in the Latin form folium has also given us the modern term folio.

When, however, the reed pen and the pencil brush and their kindred substances denominated colored liquids or inks, came into vogue, some material on which characters could be inscribed and preserved in the shape of continuous rolls for record and other uses became necessary. The papyrus plant seems to have met every requirement. It is a noteworthy fact that all information which can be derived from any source, specifically calls attention to papyrus and sometimes the inner barks of trees as being coexistent with pen and ink.

Varro has been credited with many statements which in the light of investigation and discovery are proved to be incorrect. One of these is in effect that the use of papyrus was an incident pertaining to the expeditions of Alexander the Great. This assertion is not only contradicted by Pliny, the historian, who calls attention to "books of papyrus found in the tomb of Numa " (Numa Pompilius, the second king of Rome, B. C. 716-672,) but even at this late day many monuments of ancient papyri are still extant and belonging to periods more than a thousand years before Alexander's time.

The real facts in respect to this matter are, that the introduction of the use of papyrus to nations beyond the limits of Egypt was an event that did not take place until after the reign of the first Macedonian sovereign of Egypt, Ptolemy Lagus (B. C. 323) when, in return for Greek literature, Egypt gave back her papyrus. Before this epoch the Greeks had been in the habit of employing such materials as linen, wax, bark and leaves for ordinary writing purposes, while their public records were inscribed on stone, brass, lead or other metals.

Papyrus as then introduced into those western countries was the only substance for a long period employed for literary purposes.

Parchment and vellum, which were adopted there as writing materials about two centuries later, were too costly to be used so long as papyrus was within reach.

When the use of this ancient paper had become established in the countries bordering on the Mediterranean, all the MSS. assumed the form of rolls, being rolled on cylinders of wood, ivory, bronze, glass and other substances. Sometimes, the ends were decorated by various ornaments. As a rule only one side of the material was written upon. This was due largely to the fact of its brittle character which would cause it to break if rolled or bent the wrong way.

The ancient manufacture of papyrus for export was carried on in Egypt on an extensive scale and in the most systematic manner. A gradual improvement in quality was the result, some of the kinds being given well-known Roman names which are mentioned by contemporary writers. The kind employed by the Romans for ordinary use was designated Charta. More expensive qualities were known as "Augusta," "Livinia," "Hieratica," etc., the latter being reserved for religious books. Some kinds were sold by weight and employed by the tradesmen for wrapping purposes, while the bark of the plant was manufactured into cord and rope.

The methods of the manufacture of papyrus as a writing material Pliny undertakes to describe at great length, and while he asserts many things from probable knowledge and the information at hand in his time, yet he is not always correct. He says that the reed stalks were cut into lengths and separated "by splitting the successive folds of the stalk with a fine metal point."

Mr. Knight, who investigated this matter with care, is authority for the statement, that the papyrus stalk as seen under the microscope shows that it does not possess successive folds, but is a triangular stalk with a single envelope with a pith on the inside, which could only be divided into slices with a knife, either in stripes of a width permitted by the sides of the prism, or else shaved round and round, like the operation of cork making, and producing a long spiral shaving.

In the description which Pliny gives of the various homes of this plant in Egypt, he calls particular attention to its abundance in marshy places where the Nile overflows and stagnates: "It grows like a great bulrush from fibrous, reedy roots, and runs up in several triangular stalks to a considerable height." They possessed large tufted heads, but only the stem was fit for making into paper. After the pellicles or thin coats were removed from the stalk, they were laid upon tables two or more over each other and glued together with the muddy and glutinous water of the Nile or with fine paste made of wheat flour; after being pressed and dried they were made smooth with a ruler and then rubbed over with a glass hemisphere. The size of the paper seldom exceeded two feet.

Papyrus was also known to the Hebrews.

The Prophet Isaiah (B. C. 752) refers to this plant when he says:

"The paper reeds by the brooks, and everything sown by the brooks, shall wither, be driven away and be no more."

Which prediction seems to have been long ago fulfilled as the plant is now exceedingly rare.

The manufacture of Egyptian paper from papyrus it is said was quite an industry in the ancient city of Memphis more than six hundred years before the Christian era.

The Mexicans employed for writing a paper which somewhat resembled the Egyptian papyrus. It was prepared from the aloe, called by the natives Maguey which grows wild over the tablelands of Mexico. It could be easily colored and seemed to bind to ink very closely. It could be rolled up in scrolls just like the more ancient rolls of papyrus.

The following account of an interesting discovery of a fragment of one of the "Orations of Hyperides," by Mr. Harris, the well-known Oriental scholar, is derived from the London Athenaeum:

"In the winter of 1847 Mr. Harris was sitting in his boat, under the shade of the well-known sycamore, on the western bank of the Nile, at Thebes, ready to start for Nubia, when an Arab brought him a fragment of a papyrus roll, which he ventured to open sufficiently to ascertain that it was written in the Greek language, and which he bought before proceeding further on his journey. Upon his return to Alexandria, where circumstances were more favorable to the difficult operation of unrolling a fragile papyrus, he discovered that be possessed a fragment of the oration of Hyperides against Demosthenes, in the matter of Harpalus, and also a very small fragment of another oration, the whole written in extremely legible characters, and of a form or fashion which those learned in Greek MSS. consider to be of the time of the Ptolemies. With these interesting fragments of orations of an orator so celebrated is Hyperides, of whose works nothing, is extant but a few quotations in other Greek writers, he embarked for England. Upon his arrival there he submitted the precious relics to the inspection of the Council and members of the Royal Society of Literature, who were unanimous in their judgment as to the importance and genuineness of the MSS.; and Mr. Harris immediately set to work, and with his own hand made a lithographic facsimile of each piece. Of this performance a few copies were printed and distributed among the savants of Europe,—and Mr. Harris returned to Alexandria, whence he has made more than one journey to Thebes in the hope of discovering some other portion of the volume, of which he already had a part. In the same year (1847) another English gentleman, Mr. Joseph Arden, of London, bought at Thebes a papyrus, which he likewise brought to England. Induced by the success of Mr. Harris, Mr. Arden submitted his roll to the skilful and experienced hands of Mr. Hogarth; and upon the completion of the operation of unrolling, the MSS. was discovered to be the terminating portion of the very same volume of which Mr. Harris had bought a fragment of the former part in the very same year, and probably of the very same Arabs. No doubt now existed that the volume, when entire, consisted of a collection of, or a selection from, the orations of the celebrated Athenian orator, Hyperides.

"The portion of the volume which has fallen into the possession of Mr. Arden contains 'fifteen continuous columns of the "Oration for Lycophron," to which work three of Mr. Harris's fragments appertained; and likewise the "Oration for Euxenippus," which is quite complete and in beautiful preservation. Whether, as Mr. Babington observes in his preface to the work, any more scraps of the "Oration for Lycophron" or of the "Oration against Demosthenes" remain to be discovered, either in Thebes or elsewhere, may be doubtful, but is certainly worth the inquiry of learned travellers.' The condition, however, of the fragments obtained by Mr. Harris but too significantly indicate the hopelessness of success. The scroll had evidently been more frequently rolled and unrolled in that particular part, namely, the speech of Hyperides in a matter of such peculiar interest as that involving the honor of the most celebrated orator of antiquity; it had been more read and had been more thumbed by ancient fingers than any other speech in the whole volume; and hence the terrible gap between Mr. Harris's and Mr. Arden's portions Those who are acquainted with the brittle, friable nature of a roll of papyrus in the dry climate of Thebes, after being buried two thousand years or more and then coming first into the hands of a ruthless Arab, who, perhaps, had rudely snatched it out of the sarcophagus of the mummied scribe, will well understand how dilapidations occur. It frequently happens that a single roll, or possibly an entire box, of such fragile treasures is found in the tomb of some ancient philologist or man of learning, and that the possession is immediately disputed by the company of Arabs who may have embarked on the venture. To settle the dispute, when there is not a scroll for each member of the company, an equitable division is made by dividing the papyrus and distributing the portions. Thus, in this volume of Hyperides, it seems that it has fallen into two pieces at the place where it had most usually been opened, and where, alas! it would have been most desirable to have kept it whole; and that the smaller fragments have been lost amid the dust and rubbish of the excavation, while the two extremities have been made distinct properties, which have been sold, as we have seen, to separate collectors. So, at all events, such matters are managed at Thebes.

"Mr. Harris mentions fragments of the 'Iliad,' which he had purchased of some of the Arab disturbers of the dead in the sacred cemeteries of Middle Egypt, most probably Saccara."

The oldest known specimens of the Greek papyri and which were found in Egypt, have a range of one thousand years; that is, from the third century B. C. to the seventh century A. D.

The first discovery of Greek papyri was made at Herculaneum in 1752. Papyrus, however, in the most ancient, periods was not the only pliable material used to write on which could be rolled on cylinders. Linen or cloth, which had been first treated with substances which filled the interstices and characteristic of our oil-cloth, the inner bark of certain trees, or in fact any material which would receive ink and roll around a cylinder was in vogue. This form of manuscript was later termed by the Romans rolles, to roll round, or more commonly volvere, to roll over.

It is not certain, however, that this character of manuscript immediately superseded the tablet form of records inscribed on wood or metal. Noel Humphreys is one of several to suggest:

"The reference to the 'pen of a ready writer,' mentioned in the Psalms of David (B. C. 1086- 1016) could scarcely be the sharp point, or stilus, by means of which characters were engraved upon wood or metal, but rather the calamus or juncas, used for writing with a dark fluid upon bark or linen. The word volume indeed occurs in Psalms xxxix., and these volumina or volumes must have been either rolls of leaves, or bark, or Egyptian papyrus."

Some writers like Casley, Purcelli, Haygen, Calmet, and others, who also more or less discuss this subject, do not view it entirely the same.

CHAPTER XXIX.

ANCIENT INK BACKGROUNDS (PARCHMENT AND VELLUM).

THE PERGAMUS LIBRARY COMPOSED PRINCIPALLY OF PARCHMENT VOLUMES—CAUSES WHICH CONTRIBUTED TO THE SUBSTITUTION OF PARCHMENT FOR PAPYRUS—ANECDOTE ABOUT EUMENES AND PTOLEMY PHILADELPHUS—INVENTION OF METHOD WHICH MADE SKINS AVAILABLE FOR FLUID INK WRITING—INTRODUCTION OF DRESSED SKINS THE FIRST STEP TOWARDS THE MODERN FORM OF BOOKS—WHEN PARCHMENT AND VELLUM SUPERSEDED OTHER SUBSTANCES AS A GENERAL MATERIAL FOR WRITING UPON—MANUFACTURE OF BARK PAPER PREVIOUS TO THE INTRODUCTION OF THE LINEN PAPER OF THE EAST—SOME OBSERVATIONS ABOUT CHINESE PAPER—ALLUSIONS OF CLASSICAL WRITERS TO INSCRIPTIONS ON SKINS AND DISCOVERY

OF SPECIMENS—EMPLOYMENT OF PARCHMENT BY THE HEBREWS—OLD SCRIPTURAL MSS. DISCOVERED ON PARCHMENT—NAMES OF THE MOST VALUABLE NEW TESTAMENT CODICES—STORY OF THE DISCOVERY OF THE SINAITIC CODEX AS TOLD BY MADAN—ASSERTION OF SIMONIDES THAT HE FORGED IT—PAMLIMPSESTS THE LINK BETWEEN CLASSICAL TIMES AND THE MIDDLE AGES—OBSERVATIONS ABOUT THEM AND SOME DISCOVER-IES
OF THE MORE FAMOUS ONES—USE OF PAPYRUS, PARCHMENT AND VELLUM TOGETHER IN MSS. BOOKS—OBSERVATIONS BY THOMPSON—CHARACTER OF THE ROLLS AND RECORDS BELONGING TO EARLY PARLIAMENTARY TIMES IN ENGLAND—COMPARATIVE METHODS OF THEIR PREPARATION—MODES OF DEPOSITING AND CARRYING ANCIENT ENGLISH RECORDS —METHOD OF FINDING PARTICULAR DOCUMENTS—THE INDIVIDUALS WHO HANDLED THE BOOKS OF THOSE EPOCHS—CITATIONS FROM KNIGHT'S "LIFE OF CAXTON"—REMARKS BY WARTON—EXPENSE ACCOUNT OF SIR JOHN HOWARD—METHODS OF THE TRANSCRIBERS AND LIMNERS OF THOSE TIMES—MODERN METHODS OF PREPARING PARCHMENT AND VELLUM—CITATION FROM THE PENNY CYCLOPaeDIA—PASSAGE FROM A SERMON OF THE ARCHBISHOP OF TOURS—ANECDOTE ABOUT THE COUNT OF NEVERS.

THE great abundance of papyrus in Egypt, the chief source of its supply, the genius and magnificence of the rulers of that country, and the army of learned men who resorted thither, caused it to become the principal home of those immense libraries of antiquity already mentioned as having perished by fire and tumults included in periods between B. C. 48 and A. D. 640.

The Pergamus library which was deposited by Cleopatra, B. C. 32, in the city of Alexandria, is said to have been composed almost wholly of parchment written volumes. The reason or cause of such employment, of parchment in preference to papyrus is attributed to jealousies existing between Eumenes, King of Pergamus, and Ptolemy Philadelphus, the ruler of Egypt, contemporaries of each other.

This Ptolemy, B. C. 202, issued an edict prohibiting the exportation of papyrus from Egypt, and hoped thereby to rid himself of foreign rivals in the formation of libraries; also that he might never be subject to the inconvenience of wanting paper for the multitude of scribes whom he kept constantly employed, both to write original manuscripts as well as to multiply them by duplication.

Before this period the exportation of papyrus had been a very considerable article of Egyptian commerce, but thereafter it became much curtailed, and about A. D. 950 had ceased altogether.

Eumenes, it appears, was not to be deterred from his favorite study and pastime, so lie contrived a peculiar mode of dressing skins, which seems to have answered very fully the requirements of fluid-ink writing methods and thus avoiding the necessity of employing paints, the only material which would "bind" to undressed parchment (skins).

That the refined and luxurious Romans, after the introduction of parchment, vellum, and paper, insisted on an improvement in quality and appearance is certain. This appears from various passages in their best authors. Ovid, writing to Rome from his place of exile, complains bitterly that his letter must be sent plain, simple, and without the customary embellishments.

We can safely date the first step towards the modern form of books to the introduction of dressed skins (parchment and vellum), as surfaces to receive ink writing. These materials could be formed into leaves, instead of metal, wood, ivory, or wax tablets, a use to which papyrus could not be put on account of its brittleness. Thus originated the libri quadrali, or square books, which eventually superseded the ancient volumina (rolls).

Parchment and vellum gradually superseded all other substances in Europe as a general material for writing upon, after the third or fourth century. The employment of papyrus, however, in ecclesiastical centers continued even as late as the eleventh century.

A kind of bark paper was manufactured in Europe previous to the introduction of linen ("cotton," "Bombycina") paper from the

East. The ancient Chinese made various kinds of paper and had a method of producing pieces sometimes forty feet in length. The Chinese record, called "Sou kien tchi pou," states that a kind of paper was made from hemp, and another authority (Du Halde) observes, "that old pieces of woven hemp were first made into paper in that country about A. D. 95, by a great mandarin of the palace." Linen rags were afterwards employed by the Chinese.

The introduction of "linen" paper into Europe did not materially affect or interfere with the use of parchment or vellum until after the invention of printing in the fifteenth century.

The class of substances to which parchment and vellum belong has already received some consideration but is a subject well worth some further discussion.

Allusions are found in some of the classical writers to inscriptions written on the skins of goats and sheep; it has, indeed, been asserted by some scholars that the Books of Moses were written on such skins. Dr. Buchanan many years ago discovered, in the record chest of some Hebrews at Malabar, a manuscript copy of the greater part of the Pentateuch, written in Hebrew on goat's skins. The goat skins were thirty-seven in number, dyed red, and were sewn together, so as to form a roll forty-eight feet in length by twenty-two inches in width. At what date this was written cannot be now determined, but it is supposed to be extremely ancient.

The Hebrews began, early after the invention of parchment, to write their scriptures on this material, of which the rolls of the law used in their synagogues are still composed.

Scriptural, like many other classes of MSS. originating previous to the eighth century and ink written either on parchment or vellum, or both, are in capital letters without spaces between words and exceedingly rare. The more important and valuable of them which apply to the New Testament are respectively known as the Sinaitic, the Vatican and the Alexandrian, many of whose various translations and readings are incorporated by Tischendorf in his Leipzig edition of the English New Testament. The stories relating to the discovery and obtaining of these relics of the first centuries of our era are startling ones. The reputation and standing, however, of the discoverers, and the investigations subsequently made by known scholars of their time, serves to invest them with a certain degree of truthfulness. The most interesting is the story about the Sinaitic codex, the oldest of any extant and which is best told by Madan:

"The story of the discovery of this famous manuscript of the Bible in Greek, the oldest existing of all the New Testament codexes, and in several points the most interesting, reads like a romance. Constantine Tischendorf, the well- known editor of the Greek Testament, started on his first mission litteraire in April, 1844, and in the next month found himself at the Convent of St. Catherine, at the foot of Mount Sinai. There, in the middle of the hall, as he crossed it, he saw a basket full of old parchment leaves on their way to the burning, and was told that two baskets had already gone! Looking at the leaves more closely, he perceived that they were parts of the Old Testament in Greek, written in an extremely old handwriting. He was allowed to take away forty-three leaves; but the interest of the monks was aroused, and they both stopped the burning, and also refused to part with any more of the precious fragments. Tischendorf departed, deposited the forty- three leaves in the Leipsig Library, and edited them under the title of the Codex Friderico-Au- gustanus, in compliment to the King of Saxony, in 1846. But he wisely kept the secret of their provenance, and no one followed in his track until he himself went on a second quest to the monastery in 1853. In that year he could find no traces whatever of the remains of the MSS. except a few fragments of Genesis, and returned unsuccessful and disheartened. At last, he once more took a journey to the monastery, under the patronage of the Russian Emperor, who was popular throughout the East as the protector of the Oriental Churches. Nothing could he find, however; and he had ordered his Bedouins to get ready for departure, when, happening to have taken a walk with the steward of the house, and to be invited into his room, in the course of conversation the steward said: 'I, too, have read a Septuagint,' and produced out of a wrapper of red cloth, 'a bulky kind of volume,' which turned out to be the whole of the New Testament, with the Greek text of the Epistle of Barnabas, much of which was hitherto unknown, and the greater part of the Old Testament, all parts of the very MSS. which had so long been sought! In a careless tone Tischendorf asked if he might have it in his room for further inspection, and that night (February 4-5, 1859) it 'seemed impiety to sleep.' By the next morning the Epistle of Barnabas was copied out, and a course of action was settled. Might he carry the volume to Cairo to transcribe? Yes, if the Prior's leave was obtained; but, unluckily the Prior had already started to Cairo on his way to Constantinople. By the activity of Tischendorf he was caught up at Cairo, gave the requisite permission, and a Bedonin was sent to the convent, and returned with the book in nine days. On the 24th of February, Tischendorf began to transcribe it; and when it was done, conceived the happy idea of asking for the volume as a gift to the Emperor of Russia. Probably this was the only possible plea which would have gained the main object in view, and even as it was there was great delay; but at last, on the 28th of September, the gift was formally made, and the MSS. soon after deposited in St. Petersburg, where it now lies. The date of this MSS. is supposed to be not later than A. D. 400, and has been the subject of minute inquiry in consequence of the curious statement of Simonides in 1862, that he had himself written it on Mount Athos in 1839-40."

Constantine Simonides was a Greek who was born in 1824 and is believed to have been the most versatile forger of the nineteenth century. From 1843 until 1856 he was in evidence all over Europe offering for sale fraudulent MSS. purporting to be of ancient origin.

In 1861 Madan says:

"He boldly asserted that he himself had written the whole of the Codex Sinaiticus which Tischendorf had bought in 1856 from the monastery of St. Catherine on Mount Sinai. The statement was, of course, received with the utmost incredulity; but Simionides asserted, not only that he had written it, but that, in view of the probable skepticism of the scholars, he had placed certain private signs on particular leaves of the codex. When pressed to specify these marks he gave a list of the leaves on which were to be found his initials or other monogram. The test was a fair one, and the MSS., which was at St. Petersburg, was carefully inspected. Every leaf designated by Simonides was found to be imperfect at the part where the mark was to have been found. Deliberate mutilation by an enemy, said his friends. But many thought that the wily Greek had acquired through private friends a note of some imperfect leaves in the MSS., and had made unscrupulous use of the information."

A curious kind of document, which links the classical times with the middle ages, in respect to the we of parchment, is afforded by the "palimpsests," or manuscripts from which old writing had been erased in order to make way for new. A well-prepared leaf of parchment was so costly an article in the middle ages, that the transcribers who were employed by the monastic establishments in writing often availed themselves of some old manuscript, from which they scraped off the writing; such a doubly-used piece of parchment was called a "palimpsest." This practice seems to have been followed long before, but not to so great an extent as about the fourteenth and fifteenth centuries, at which time there were persons regularly employed as "parchment-restorers." The transcribers had a regular kind of knife, with which they scratched out the old writing, and they rubbed the surface with powdered pumice stone, to prepare it for receiving the new ink. So common was this practice that when one of the emperors of Germany established the office of imperial notary, it was one of the articles or conditions attached to the holding of the office that the notary should not use "scraped vellum" in drawing deeds. Sometimes the original writing, by a careful treatment of the parchment, has been so far restored as to be visible, and it is found to be parallel, diagonal, and sometimes at right angles to the writing afterwards introduced. In many cases the ancient writing restored beneath is found to be infinitely more valuable than the monkish legends written afterwards.

Cicero's De Republica was discovered by Angelo Mai in the Vatican library written under a commentary of St. Augustine on the Psalms; and the Institutions of Gains, in the library of the chapter of Verona, were deciphered in like manner under the works of St. Jerome.

Papyrus, parchment, and vellum were sometimes used together in the MSS. books. Thompson, author of "Greek and Latin Palaeography," observes:

"Examples, made up in book form, sometimes with a few vellum leaves incorporated to give stability, are found in different libraries of Europe. They are: The Homilies of St. Avitus, of the 6th century, at Paris; Sermons and Epistles of St. Augustine, of the 6th or 7th century, at Paris and Genoa; works of Hilary, of the 6th century, at

Vienna; fragments of the Digests, of the 6th century, at Pommersfeld; the Antiquities of Josephus, of the 7th century, at Milan; an Isidore, of the 7th century, at St. Gall. At Munich, also, is the register of the Church of Ravenna, written on this material in the 10th century."

The rolls and records connected with the early parliamentary and legal proceedings in England furnish interesting examples of the use of parchment in writing. The "Records," so often alluded to in such matters, are statements or details, written upon rolls of parchment, of the proceedings in those higher courts of law which are distinguished as "Courts of Record." It has been stated that "our stores of public records are justly reckoned to excel in age, beauty, correctness, and authority whatever the choicest archives abroad can boast of the like sort."

The records are generally made of several skins or sheets of parchment or vellum, each sheet being about three feet long and often nine to fourteen inches in width. They are either all fastened together at one end, so as to form a kind of book, or are stitched end to end, so as to constitute an extended roll. These two methods appear each to have had its particular advantages, according to the way in which, and the time at which, the manuscript was filled up. Some of the records of the former of these two kinds contain so many skins of parchment that they form a huge roll equal in size to a large bass drum, and requiring the strength of two men to lift them. Some of these on the continuous plan are also said to be of immense size; one, of modern date, is nine hundred feet in length and employs a man three hours to unroll it. The invaluable old record, known by the name of "Doomsday Book," is shaped like a book, and is much more convenient to open than most of the others. Various other legal documents, to an immense amount, are "filed," or fastened together by a string passing through them.

It seems a very strange contradiction, but it is positively asserted as a fact, that the parchment employed for these records was of very fine quality down to the time of Elizabeth, but that it gradually deteriorated afterwards, insomuch that the latest are the worst. Some of these records and rolls are written in Latin, some in Norman French, and some in English.

The modes of depositing and carrying the ancient records were curious, and there seems to have been no very definite arrangement in this respect. Great numbers were kept in pouches or bags made of leather, canvas, cordovan, or buckram; they were tied like modern reticules. When such pouches have escaped damp they have preserved the parchment records for centuries perfectly clean and uninjured. Another kind of receptacle for records was a small turned box, called a "skippet," and another was the "hanaper," or hamper, a basket made of twigs or wicker-work. Chests, coffers, and cases of various shapes and sizes formed other receptacles for the records. The mode of finding the particular document required was not by a system of paging and an index, as in a modern book, because the arrangement of the written sheets did not admit of this, but there were letters, signs, and inscriptions, or labels for this purpose; they constitute an odd assemblage, comprising ships, scales, balances, castles, plants, animals, etc.; in most instances the signs or symbols bear some analogy, or supposed analogy, with the subject of the record, such as an oak on a record relating to the forest laws, a head in a cowl on one relating to a monastery, scales on one relating to coining, etc.

At a time when books were prepared by hand instead of by printing, and when each copy became very valuable, books were treated with a degree of respect which can be hardly understood at the present day. The clergy and the monks were almost exclusively the readers of those days, and they held the other classes of society in such contempt, in all that regarded literature and learning, that Bishop de Burg, who wrote about five centuries ago, expresses an opinion that "Laymen, to whom it matters not whether they look at a book turned wrong side upwards or spread before them in natural order, are altogether unworthy of any communion with books."

It is stated by Mr. Knight, in his "Life of Caxton:"

"We have abundant evidence, whatever be the scarcity of books as compared with the growth of scholarship, that the ecclesiastics laboured most diligently to multiply books for their own establishments. In every great abbey there was a room called the Scriptorium, where boys and novices were constantly employed in multiplying the service- books of the choir, and the less valuable books for the library; whilst the monks themselves laboured in their cells upon bibles and missals. Equal pains were taken in providing books for those who received a liberal education in collegiate establishments."

Warton says:

"At the foundation of Winchester College, one or more transcribers were hired and employed by the founder to make books for the library. They transcribed and took their food within the college, as appears by computation of expenses on their account now remaining. But there are many indications that even kings and nobles had not the advantage of scholars by profession, and, possessing few books of their own, had sometimes to borrow of their more favoured subjects."

We learn from another source that the great not only procured books by purchase, but employed transcribers to make them for their libraries. The manuscript expense account of Sir John Howard, afterwards Duke of Norfolk, shows in 1467, Thomas Lympnor, that is Thomas the Limner of Bury, was paid the sum of fifty shillings and two pence for a book which he had transcribed and ornamented, including the vellum and binding. The limner's bill is made up of a number of items, "for whole vignettes, and half-vignettes, and capital letters, and flourishing and plain writing."

These transcribers and limners worked principally upon parchment and vellum, for the use of paper was by no means extensive until the invention of the art of printing. Some of the old manuscripts contain drawings representing a copier or transcriber at work, where the monk is represented as provided with a singular and tolerably complete set of apparatus to aid him in his work. The desk for containing the sheet or skin on which he is writing, the clasp to keep this sheet flat, the inkstand, the pen, and the knife, the manuscript from which the copy is being made, the desk for containing that manuscript, and the weight for keeping it in its place,—all are shown, with a clearness which, despite of bad perspective, renders them quite intelligible.

Of the two substances, parchment and vellum, before the invention of paper, another word or two may be said. Parchment is made from the skin of sheep or lambs; vellum, from that of very young calves (sometimes unborn ones), but the process of preparing is pretty much the same in both cases. When the hair or wool has been removed, the skin is steeped in lime water, and then stretched on a square frame in a light manner. While so stretched, it is scraped on the flesh side with a blunt iron, wetted with a moist rag, covered with pounded chalk, and rubbed well with pumice stone. After a time, these operations are repeated, but without the use of chalk; the skin is then turned, and scraped on the hair side once only; the flesh side is then scraped once more, and again rubbed over with chalk, which is brushed off with a piece of lambskin retaining the wool. All this is done by the skinner, who allows the skin to dry on a frame, and then cuts it out and sends it to the parchment maker, who repeats the operation with a sharper tool, using a sack stuffed with flocks (wool or hair) to lay the skin upon, instead of stretching it on a frame.

Respecting the quality, value, and preparation of parchment in past ages, it is stated in the "Penny Cyclopaedia" that parchment from the seventh to the tenth century was "white and good, and at the earliest of these periods it appears to have nearly superseded papyrus, which was brittle and more perishable. A very few books of the seventh century have leaves of parchment and papyrus mixed, that the former costly material might strengthen and support the friable paper. About the eleventh century it grew worse, and a dirty colored parchment is evidence of a want of antiquity. This may possibly arise from the circumstances that writers of this time prepared their own parchment, and they were probably not so skilled as manufacturers. A curious passage from a sermon of Hildebert, Archbishop of Tours, who was born in 1054, is a voucher for this fact. The sermon is on the "Book of Life," which he recommends his hearers to obtain:

'Do you know what a writer does? He first cleanses his parchment from the grease, and takes off the principal part of the dirt; then he entirely rubs off the hair and fibres with pumice stone; if he did not do so, the letters written upon it would not be good, nor would they last long. He then rules lines that the writing may be straight. All these things you ought to do, if you wish to possess the book which I have been displaying to you.'

At this time parchment was a very costly material. We find it mentioned that Gui, Count of Nevers, having sent a valuable present of plate to the Chartreux of Paris, the unostentatious monks returned it with a request that he would send them parchment instead."

CHAPTER XXX.

MODERN INK BACKGROUNDS (TRUE PAPER).

WHEN IT WAS THAT TRUE PAPER WAS INVENTED—CITATIONS FROM MUNSELL ABOUT CHINESE AND OTHER ANCIENT PAPER—A SHORT CHRONOLOGY FROM THE SAME AUTHOR—LINEN PAPER IN USE IN THE TWELFTH CENTURY—BOMBYCINE PAPER—DEVELOPMENTS OF THE MICROSCOPE—METHODS EMPLOYED IN ASCERTAINING ORIGIN OF LINEN PAPER BY MEERMAN—SOME OBSERVATIONS RELATIVE TO THE EVOLUTION OF PAPER —RAPID IMPROVEMENT IN QUALITY AFTER INVENTION OF PRINTING—CURIOUS CUSTOMS IN THE USE OF THE WATER MARK—NO DISTINCTIONS IN QUALITY OF PAPER USED FOR MSS. OR OTHER BOOKS—ANECDOTES AND OBSERVATIONS ABOUT THE WATER MARK—ITS VALUE IN DETECTING FRAUDS—INTERESTING ANECDOTE OF ITS USE IN FABRICATING A FRAUD—FULLER'S CHARACTERIZATION OF THE PAPERS OF DIFFERENT COUNTRIES—WHEN THE FIRST PAPER MILL WAS ESTABLISHED IN EUROPE FOR THE MANUFACTURE OF LINEN PAPER—DATE OF THE ESTABLISHMENT OF THE FIRST PAPER MILL IN AMERICA—WHO FIRST SUGGESTED WOOD AS A MATERIAL FOR MAKING PAPER—SOME NAMES OF AUTHORS ON THE SUBJECT OF PAPER—STORY OF RAG PAPER INSTRUCTIVE AS WELL AS INTERESTING.

WHEN it was that the great change occurred and true paper made of fibrous matter or rags reduced to a pulp in water was invented has been a subject of considerable thought and investigation. Munsell, in his "Chronology of Paper and Paper-Making," credits it to the Chinese, and estimates its date to be included in the first century of the Christian era. He observes:

"The Chinese paper is commonly supposed to be made of silk; but this is a mistake. Silk by itself cannot be reduced to a pulp suitable for making paper. Refuse silk is said to be occasionally used with other ingredients, but the greater part of the Chinese paper is made from the inner bark of the bamboo and mulberry tree, called by them the paper tree, hempen rags, etc. The latter are prepared for paper by being cut and well washed in tanks. They are then bleached and dried; in twelve days they are converted into a pulp, which is then made into balls of about four pounds weight. These are afterwards saturated with water, and made into paper on a frame of fine reeds; and are dried by being pressed under large stones. A second drying operation is performed by plastering the sheets on the walls of a room. The sheets are then coated with gum size, and polished with stones. They also make paper from cotton and linen rags, and a coarse yellow sort from rice straw, which is used for wrapping. They are enabled to make sheets of a large size, the mould on which the pulp is made into paper being sometimes ten or twelve feet long and very wide, and managed by means of Pulleys.

"The Japanese prepare paper from the mulberry as follows: in the month of December the twigs are cut into lengths not exceeding thirty inches and put together in bundles. These fagots are then placed upright in a large vessel containing alkaline ley, and boiled till the bark shrinks so as to allow about a half an inch of the wood to appear free at the top. After they are thus boiled they are exposed to a cool atmosphere, and laid away for future use. When a sufficient quantity has been thus collected, it is soaked in water three or four days, when a blackish skin which covered it is scraped off. At the same time also the stronger bark which is of a full year's growth is separated from the thinner, which covered the younger branches, and which yields the best and whitest paper. After it has been sufficiently cleansed out and separated, it must be boiled in clear ley, and if stirred frequently it soon becomes of a suitable nature.

"It is then washed, a process requiring much attention and great skill and judgment; for if it be not washed long enough, although strong and of good body, will be coarse and of little value; if washed too long it will afford a white paper, but will be spongy and unfit for writing upon. Having been washed until it becomes a soft and woolly pulp, it is spread upon a table and beat fine with a mallet. It is then put into a tub with an infusion of rice and breni root, when the whole is stirred until the ingredients are thoroughly mixed in a mass of proper consistence. The moulds on which sheets are formed are made of reeds cut into narrow strips instead of wire, and the process of dipping is like that of other countries. After being allowed to remain a short time in heaps under a slight pressure, the sheets are exposed to the sun, by which they are properly dried.

"The Arabians in the seventh century appear to have either discovered or to have learned from the Chinese or Hindoos, quite likely from the latter, the art of making paper from cotton; for it is known that a manufactory of such paper was established at Samarcand about the year 706 A. D, The Arabians seem to have carried the art to Spain, and to have there made paper from linen and hemp as well as from cotton.

"The art of manufacturing paper from cotton is supposed to have found its way into Europe in the eleventh century. The first paper of that kind was made of raw cotton; but its manufacture was by the Arabians extended to old worn-out cotton, and even to the smallest pieces it is said. But as there are cotton plants of various kinds, it was natural that they should produce papers of different qualities; and it was impossible to unite their woolly particles so firmly as to form a strong substantial paper, for want of sufficient skill and proper machinery, using as they did mortars and rude horse-mills. The Greeks, it is said, made use of cotton paper before the Latins. It came into Germany through Venice and was called Greek parchment.

"The Moors, who were the paper-makers of Spain, having been expelled by the Spaniards, the latter, acquainted with water mills, improved the manufacture so as to produce a paper from cotton nearly equal to that made of linen rags."

A chronology of paper relating to the earliest specimens of them can also be found in Munsell's work on that subject; several are here cited:

"A. D. 704. The Arabians are supposed to have acquired the knowledge of making paper of cotton, by their conquests in Tartary.

"A. D. 706. Casiri, a Spanish author, attributes the invention of cotton paper to Joseph Amru, in this year, at Mecca; but it is well known that the Chinese and Persians were acquainted with its manufacture before this period.

"A. D. 900. The bulls of the popes in the eighth and ninth centuries were written upon cotton paper.

"A. D. 900. Montfaucon, who on account of his diligence and the extent of his researches is great authority, wrote a dissertation to prove that charta bombycine, cotton paper, was discovered in the empire of the east toward the end of the ninth or beginning of the tenth century.

"A. D. 1007. The plenarium, or inventory, of the treasure of the church of Sandersheim, is written upon paper of cotton, bearing this date.

"A. D. 1049. The oldest manuscript in England written upon cotton paper, is in the Bodleian collection of the British Museum, having this date.

"A. D. 1050. The most ancient manuscript on cotton paper, that has been discovered in the Royal Library at Paris having a date, bears record of this year.

"A. D. 1085. The Christian successors of Moorish paper-makers at Toledo in Spain, worked the paper-mills to better advantage than their predecessors. Instead of manufacturing paper of raw cotton, which is easily recognized by its yellowness and brittleness, they made it of rags, in moulds through which the water ran off; for this reason it was called parchment cloth.

"A. D. 1100. The Aphorisms of Hippocrates, in Arabia, the manuscript of which bears this date, has been pronounced the oldest specimen of linen paper that has come to light.

"A. D. 1100. Arabic manuscripts were at this time written on satin paper, and embellished with a quantity of ornamental work, painted in such gay and resplendent colors that the reader might behold his face reflected as if from a mirror.

"A. D. 1100. There was a diploma of Roger, king of Sicily, dated 1145, in which be says that he had renewed on parchment a charter that had been written on cotton paper in 1100.

"A. D. 1102. The king of Sicily appears to have accorded a diploma to an ancient family of paper-makers who had established a manufactory in that island, where cotton was indigenous, and this has been thought to point to the origin of cotton paper, quite erroneously.

"A. D. 1120. Peter the Venerable, abbot of Clum, who flourished about this time, declared that paper from linen rags was in use in his day.

"A. D. 1150. Edrisi, who wrote at this time, tells us that the paper made at Xativa, an ancient city of Valencia, was excellent, and was exported to countries east and west.

"A. D. 1151. An Arabian author certifies that very fine white cotton paper was manufactured in Spain, and Cacim aben Hegi assures us that the best was made at Xativa. The Spaniards being acquainted with water-mills, improved upon the Moorish method of grinding the raw cotton and rags; and by stamping the latter in the mill, they produced a better pulp than from raw cotton, by which various sorts of paper were manufactured, nearly equal to those made from linen rags.

"A. D. 1153. Petrus Mauritius (the Abbi de Cluni), who died in this year, has the following passage on paper in his Treatise against the Jews; 'The books we read every day are made of sheep, goat, or calf skin; or of rags (ex rasauris veterum pannorum),' supposed to allude to modern paper.

"A. D. 1178. A treaty of peace between the kings of Aragon and Castile is the oldest specimen of linen paper used in Spain with a date. It is supposed that the Moors, on their settlement in Spain, where cotton was scarce, made paper of hemp and flax. The inventor of linen-rag paper, whoever he was, is entitled to the gratitude of posterity.

"A. D. 1200. Casiri positively affirms that there are manuscripts in the Escurial palace near Madrid, upon both cotton and hemp paper, written prior to this time."

Abdollatiph, an Arabian physician, who visited Egypt in 1200, says that the linen mummy-cloths were habitually used to make wrapping paper for the shopkeepers.

A document with the seals preserved dated A. D. 1239 and signed by Adolphus, count of Schaumburg is written on linen paper. It is preserved in the university of Rinteln, Germany, and establishes the fact that linen paper was already in use in Germany.

Specimens of flax paper and still extant are quite numerous, a very few of them having dates included in the eighth and ninth centuries.

The charta Damascena, so-called from the fact of its manufacture in the city of Damascus, was in use in the eighth century. Many Arabian MSS. on such a paper exist dating from the ninth century.

The charta bombycina (bombyx, a silk and cotton paper) was much employed during mediaeval periods.

The microscope, however, has demonstrated conclusively many things formerly in doubt and relating particularly to the matter of the character of fibre used in paper-making. One of the most important is the now established fact that there is no difference between the fibres of the old cotton and linen papers, as made from rags so named.

To ascertain the precise period and the particular nation of Europe, when and among whom the use of our common paper fabricated from linen rags first originated, was a very earnest object of research with the learned Meerman, author of a now exceedingly rare work on this subject and published in 1767. His mode of inquiry was unique. He proposed a reward of twenty-five golden ducats, to whoever should discover what on due examination should appear to be the most ancient manuscript or public document inscribed on paper manufactured from linen rags. This proposal was distributed through all parts of Europe. His little volume contains the replies which Meerman received. The scholars who remitted the result of their investigations were unable to distinguish between what they estimated as cotton or linen rags. They did, however, establish the fact that paper made of linen rags existed before 1308, and some of them even sought to give the honor of the invention to Germany. They also asserted that the most ancient English specimen of such a paper belonged to the year 1342.

The transformation of paper made from every conceivable fibrous material into what is commonly known as "linen" or true paper was of slow growth until after the invention of printing. Following that great event it is surprising, how, in so short a period, the manufacturers of paper improved its quality and the degree of excellence which it later attained. They imitated the old vellum so closely that it was even called vellum and is so known to this day. This class of paper was employed both for writing and printing purposes and has never been excelled, surpassing any like productions of modern times.

A curious custom came into vogue during the early infancy of the "linen" paper industry, which is of so much interest and possesses so curious a history as to be well worth mentioning. It is the water mark as it is commonly but erroneously termed in connection with paper manufacture.

Its origin dates back to the thirteenth century, though the monuments indicating its use before the time of printing are but few in number.

The real employment of the water mark may be said to have commenced at the time when it was a custom of the first printers to omit their names from their works. Also, it is to be considered that at this period comparatively few people could either read or write and therefore pictures, designs or other marks were employed to enable them to distinguish the paper of one manufacturer from another. These marks as they became common naturally gave their names to the different sorts of paper.

The earliest known water mark on linen paper represented a picture of a tower and was of the date of 1293. The next known water mark which can be designated is a ram's head and is found in a book of accounts belonging to an official of Bordeaux which was then subject to England. It is dated 1330.

In the fifteenth century there were no distinctions in the quality of paper used for manuscripts or for books. In the Mentz Bible of 1462 are to be found no less than three sorts of paper. Of this Bible, the water mark in some sheets is a bull's head simply, and in others a bull's head from whose forehead rises a long line, at the end of which is a cross. In other sheets the water mark is a bunch of grapes.

In 1498 the water mark of paper consisted of an eight pointed star within a double circle. The design of an open hand with a star at the top which was in use as early as 1530, probably gave the name to what is still called hand paper.

It appears that even so high a personage as Henry VIII of England in 1540 utilized the water mark in order to show his contempt for and animosity to Pope Paul III, with whom he had then quarreled, gave orders for the preparation of paper, the water mark of which was a hog with a miter: this he used for his private correspondence.

A little later, about the middle of the sixteenth century, the favorite paper mark was the jug or pot, from which would appear to have originated the term pot paper. Still another belonging to this period was the device of a glove.

At the beginning of the seventeenth century, the device was a fool's cap and which has continued by name as the particular size which we now designate fool's cap.

The water mark has continued to increase in popularity and today may be found in almost any kind of paper, either in the shape of designs, figures, numbers or names.

The circumstance of the water mark has at various times been the means of detecting frauds, forgeries and impositions in our courts of law and elsewhere. The following is introduced as a whimsical example of such detections and is said to have occurred in the fifteenth century, and is related by Beloe, London, 1807:

"The monks of a certain monastery at Messina exhibited to a visitor with great triumph, a letter which they claimed had been written in ink by the Virgin Mary with her own hand, not on the ancient papyrus, but on paper made of rags. The visitor to whom it was shown observed with affected solemnity, that the letter involved also a miracle because the paper on which it was written could not have been in existence until over a thousand years after her death."

An interesting example of the use of water marks on paper for fraudulent purposes is to be found in a pamphlet entitled "Ireland's Confessions." This person, a son of Samuel Ireland, who was a distinguished draughtsman and engraver, about the end of the eighteenth century fabricated a pretended Shakespeare MSS., which as a literary forgery was the most remarkable of its time. Previous to his confessions it had been accepted by the Shakespearean scholars as unquestionably the work of the immortal bard. The following is a citation from his Confessions:

"Being thus urged forward to the production of more manuscripts, it became necessary that I should posses; a sufficient quantity of old paper to enable me to proceed; in consequence of which I applied to a book-seller named Verey, in Great May's buildings, St. Martin's Lane, who, for the sum of five shillings, suffered me to take from all the folio and quarto volumes in his shop the fly leaves which they contained. By this means I was amply stored with that commodity—nor did I fear any mention of the circumstance by Mr. Verey, whose quiet, unsuspecting disposition, I was well convinced, would never lead him to make the transaction public; in addition to which, he was not likely even to know anything concerning the supposed Shakespearean discovery by myself, and even if he had, I do not imagine that my purchase of the old paper in question would have excited in him the smallest degree of suspicion. As I was fully aware, from the variety of water-marks, which are in existence at the present day, that they must have constantly been altered since the period of Elizabeth and being for some time wholly unacquainted with the water-marks of that age, I very carefully produced my first specimens of the writing on such sheets of old paper as had no marks whatever. Having heard it frequently stated that the appearance of such marks on the papers would have greatly tended to establish their validity, I listened attentively to every remark which was made upon the subject, and from thence I at length gleaned the intelligence that a jug was the prevalent water-mark of the reign of Elizabeth; in consequence of which I inspected all the sheets of old paper then in my possession, and having selected such as had the jug upon them, I produced the succeeding manuscripts upon these, being careful, however, to mingle with them a certain number of blank leaves, that the production on a sudden of so many water-marks might not excite suspicion in the breasts of those persons who were most conversant with the manuscripts."

Fuller, writing in 1662, characterizes the paper of his day:

"Paper participates in some sort of the character of the country which makes it; the Venetian being neat, subtle, and court-like; the French light, slight, and slender; and the Dutch thick, corpulent, and gross, sticking up the ink with the sponginess thereof. And he complains of the 'vast sums of money expended in our land for paper out of Italy, France, and Germany, which might be lessened were it made in our nation.' "

Ulman Strother in 1390 started his paper mill at Nuremberg in Bavaria which was the first paper mill known to have been established in Germany, and is said to have been the only one in Europe then manufacturing paper from linen rags.

Among the privy expenses of Henry VII of the year 1498 appears the following entry: "A reward given to the paper mill, 16s. 8d." This is probably the paper mill mentioned by Wynkin de Worde, the father of English typography. It was located at Hertford, and the water mark he employed was a star within a double circle.

The manufacture of paper in England previous to the revolution of 1688 was an industry of very small proportions, most of the paper being imported from Holland.

The first paper mill established in America was by William Rittenhouse who emigrated from Holland and settled in Germantown, Pa., in 1690. At Roxborough, near Philadelphia, on a stream afterwards called Paper Mill run, which empties into the Wissahicken river, was located the site which in company with William Bradford, a printer, he chose for his mill. The paper was made from linen rags, mostly the product of flax raised in the vicinity and made first into wearing apparel.

It was Reaumer, who in 1719 first suggested the possibility of paper being made from wood. He obtained his information on this subject from examination of wasps' nests.

Matthias Koops in 1800 published a work on "Paper" made from straw, wood and other substances. His second edition appeared in

1801 and was composed of old paper re-made into new. Another work on the subject of "Paper from Straw, &c.," by Piette, appeared in 1835, which said work contains more than a hundred pages, each one of which was made from a different kind of material.

Many other valuable works are obtainable which treat of rag paper manufacture and the stories they tell are instructive as well as interesting.

CHAPTER XXXI.

MODERN INK BACKGROUNDS (WOOD PAPER AND "SAFETY" PAPER).

SOME GENERAL OBSERVATIONS ABOUT PAPER-MAKING MATERIALS—PROBABILITIES AS TO THE FUTURE OF THE PUBLIC RECORDS—ESTIMATION OF SUCH MATTERS BY THE LATE POPE—INVENTION OF WOOD-PULP PAPER —ITS LASTING QUALITIES—THE THREE KINDS OF SUCH PAPER DEFINED—DISCUSSION OF THE SUBJECT OF FUNGI IN PAPER BY GLYDE—SOME TESTS TO ASCERTAIN THE MATERIAL OF WHICH PAPER IS COMPOSED— TESTS AS TO SIZING AND THE DETERMINATION OF THE DIRECTION OF THE GRAIN—ABSORBING POWERS OF BLOTTING PAPER—TESTS FOR GROUND WOOD—NEW MODE OF ANALYSTS—WHEN THE FIRST "SAFETY" PAPER WAS INVENTED—THE MANY KINDS OF "SAFETY" PAPER AND PROCESSES IN THEIR MANUFACTURE— CHRONOLOGICAL REVIEW COVERING THIS SUBJECT— SURVEY OF THE VARIOUS PROCESSES IN THE TREATMENT AND USE OF "SAFETY" PAPER—ONLY THREE CHEMICAL "SAFETY" PAPERS NOW ON THE MARKET— WHY IT IS POSSIBLE TO RAISE SOME MONETARY INSTRUMENTS.

PAPER manufacturers have tried all the pulp-making substances. This statement to the unlearned must seem curious, because in the very early times they were content with a single material and that did not even require to be first made into the form of pulp. When the supply of papyrus failed, it was rags which they substituted. By the simplest processes they produced a paper with which our best cannot compare. In some countries great care is exercised in selecting the quality of paper for official use, in others none at all.

What will be the state of our archives a few hundred years hence, if they be not continually recopied?

Some of the printed paper rots even more quickly than written.

The late Pope at one time invited many of the savants, chemists and librarians of Europe, to meet at Einsiedlen Abbey in Switzerland. He requested that the subject of their discussions should be both ink and paper. He volunteered the information, already known to the initiated, that the records of this generation in his custody and under his control were fast disappearing and unless the writing materials were much improved he estimated that they would entirely disappear. It is stated that at this meeting the Pope's representative submitted a number of documents from the Vatican archives which are scarcely decipherable though dated in the nineteenth century. In a few of those of dates later than 1873 the paper was so tender that unless handled with exceptional care, it would break in pieces like scorched paper.

These conditions are in line with many of those which prevail with few exceptions in every country, town or hamlet.

A contributory cause as we know is a class of poor and cheap inks now in almost universal use. The other is the so-called "modern" or wood-pulp paper in general vogue.

Reaumur, as already stated, back in 1719 suggested from information gathered in examinations of wasps' nests, that a paper might be manufactured from wood. This idea does not appear to have been acted upon until many years later, although in the interim inventors were exhausting their ingenuity in the selection of fibrous materials from which paper might be manufactured.

The successful introduction of wood as a substitute for or with rags in paper manufacture until about 1870 was of slow growth; since which time vast quantities have been employed. In this country alone millions of tons of raw material are being imported to say nothing of home products.

Its value in the cause of progress of some arts which contribute greatly to our comfort and civilization cannot be overestimated, but nevertheless the wood paper is bound to disintegrate and decay, and the time not very far distant either. Hence, its use for records of any kind is always to be condemned.

There are three classes of wood pulp; mechanical wood, soda process, and the sulphite. The first or mechanical wood is a German invention of 1844, where the logs after being cut up into proper blocks, were then ground against a moving millstone against which they were pressed and with the aid of flowing water reduced to a pulpy form. This pulp was transported into suitable tanks and then pumped to the "beaters."

The soda process wood and sulphite wood pulp are both made by chemical processes. The first was invented by Meliner in 1865. The preparation of pulp by this process consists briefly in first cutting up the logs into suitable sections and throwing them into a chipping machine. The chips are then introduced into tanks containing a strong solution of caustic soda and boiled under pressure.

The sulphite process is substantially the same except that the chips are thrown into what are called digesters and fed with the chemicals which form an acid sulphite. The real inventor of this latter process is not known.

The chemicals employed in both of these processes compel a separation of the resinous matters from the cell tissues or cellulose. These products are then treated in the manufacturing of paper with few variations, the same as the ordinary rag pulp.

These now perfected processes are the results of long and continuing experimentations made by many inventors.

The following paper was read before the London Society of Arts by Mr. Alfred Glyde, in May, 1850, and is equally applicable to some of the wood paper of the present day:

"Owing to the imperfections formerly existing in the microscope, little was known of the real nature of the plants called fungi until within the last few years, but since the improvements in that instrument the subject of the development, growth, and offices of the fungi has received much attention. They compose, with the algae and lichens, the class of thallogens (Lindley), the algae existing in water, the other two in air only. A fungus is a cellular flowerless plant, fructifying solely by spores, by which it is propagated, and the methods of attachment of which are singularly various and beautiful. The fungi differs from the lichens and algae in deriving their nourishment from the substances on which they grow, instead of from the media in which they live. They contain a larger quantity of nitrogen in their constitution than vegetables generally do, and the substance called 'fungine' has a near resemblance to animal matter. Their spores are inconceivably numerous and minute, and are diffused very widely, developing themselves wherever they find organic matter in a fit state. The principal conditions required for their growth are moisture, heat, and the presence of oxygen and electricity. No decomposition or development of fungi takes place in dry organic matter, a fact illustrated by the high state of preservation in which timber has been found after the lapse of centuries, as well as by the condition of mummy-cases, bandages, etc., kept dry in the hot climate of Egypt. Decay will not take place in a temperature below that of the freezing point of water, nor without oxygen, by excluding which, is contained in the air, meat and vegetables may be kept fresh and sweet for many years.

"The action which takes place when moist vegetable substances are exposed to oxygen is that of slow combustion ('eremacausis'), the oxygen uniting with the wood and liberating a volume of carbonic acid equal to itself, and another portion combining with the hydrogen of the wood to form water. Decomposition takes place on contact with a body already undergoing the same change, in the same manner that yeast causes fermentation. Animal matter enters into combination with oxygen in precisely the same way as vegetable matter, but as, in addition to carbon and hydrogen, it contains nitrogen, the products of the eremacausis are more numerous, being carbon and nitrate of ammonia, carburetted and sulphuretted hydrogen, and water, and these ammoniacal salts greatly favor the growth of fungi. Now paper consists essentially of woody fibre, having animal matter as size on its surface. The first microscopic symptom of decay in paper is irregularity of surface, with a slight change of color, indicating the commencement of the process just noticed, during which, in addition to carbonic acid, certain organic acids are formed, as crenic and ulmic acids, which, if the paper has been stained by a coloring matter, will form spots of red on the surface. The same process of decay goes on in parchment as in paper, only with more rapidity, from the presence of nitrogen in its composition. When this decay has begun to take place, fungi are produced, the most common species being Penicilium glaucum. They insinuate themselves between the fibre, causing a freer admission of air, and consequently hasten the decay. The substances most successfully used as preventives of decay are the salts of mercury, copper, and zinc. Bichloride of mercury (corrosive sublimate) is the material employed in the kyanization of timber, the probable mode of action being its combination with the albumen of the wood, to form an insoluble compound not susceptible of spontaneous decomposition, and therefore incapable of exciting fermentation. The antiseptic power of corrosive sublimate may be easily tested by mixing a little of it with flour paste, the decay of which, and the appearance of fungi, are quite prevented by it. Next to corrosive sublimate in antiseptic value stand the salts of copper and zinc. For use in the preservation of paper the sulphate of zinc is better than the chloride, which is to a certain extent delinquescent."

There are numerous paper tests which include the matter of sizing, direction of the grain, absorbing powers, character of ingredients, etc. A few of them are cited.

SIZING. — The everyday tests as to hardness of sizing answer every ordinary purpose: Moisten with the tongue, and if the paper is slack-sized you can detect it often by the instant drawing or absorption of the moisture. Watch the spot moistened, and the longer it remains wet the better the paper is sized. Look through the spot dampened — the poorer the sizing the more transparent is the paper where it is wet. If thoroughly sized no difference will be apparent between the spot dampened and the balance of the sheet. When there is a question as to whether a paper is tub or engine sized, it can be usually decided by wetting the forefinger and thumb and pressing the sheet between them. If tub-sized, the glue which is applied to the surface will perceptibly cling to the fingers.

TO TEST THE INK RESISTING QUALITY OF PAPER. — Draw a heavy ink line across the sheet. If the paper is poorly sized, a feathery edge will appear, caused by spreading of the ink. Slack-sized paper will be penetrated by the ink, which will plainly appear on the reverse side of the sheet.

TO DETERMINE THE DIRECTION OF THE GRAIN.— An easy but sure test to determine the direction of the grain in a sheet of paper, which will be found useful and worth remembering, is as follows:

For instance, the size of sheet is 17x22 inches. Cut out a circular piece as nearly round as the eye can judge; before entirely detaching from the sheet, mark on the circle the 17-inch way and the 22-inch way; then float the cut out piece on water for a few seconds; then place on the palm of the hand, taking care not to let the edges stick to the hand, and the paper will curl until it forms a cone; the grain of the paper runs the opposite way from which the paper curls.

ABSORBING POWERS OF BLOTTING PAPER.—Comparative tests as to absorbing powers of blotting can be made between sheets of same weight per ream by allowing the pointed corner of a sheet to touch the surface of a drop of ink. Repeat with each sheet to be tested, and compare the height in each to which the ink has been absorbed. A well-made blotting paper should have little or no free fibre dust to fill with ink and smear the paper.

TEST FOR GROUND WOOD.—Make a streak across the paper with a solution of aniline sulphate or with concentrated nitric acid; the first will turn ground wood yellow, the second will turn it brown. I give aniline sulphate the preference, as nitric acid acts upon unbleached sulphite, if present in the paper, the same as it acts upon ground wood, viz., turning it brown.

Phloroglucin gives a rose-red stain on paper containing (sulphite) wood pulp, after the specimen has been previously treated with a weak solution of hydrochloric acid.

About the end of the eighteenth century it became necessary to make special papers denominated "safety paper." Their manufacture has continued until the present day although much limited, largely because of the employment of mechanical devices which seek to safety monetary instruments. Such safety papers are of several kinds.

1. Paper made with distinguishing marks to indicate proprietorship, as with the Bank of England water mark, to imitate which is a felony. Or the paper of the United States currency, which has silk fibers united with the pulp, the imitation of which is a felony.

2. Paper made with layers or materials which are disturbed by erasure or chemical discharge of written or printed contents, so as to prevent fraudulent tampering.

3. Paper made of peculiar materials or color, to prevent copying by photographic means.

A number of processes may be cited:

One kind is made of a pulp tinged with a stain easily affected by chlorine, acids, or alkalis, and is made into sheets as usual.

Water marks made by wires twined among the meshes of the wire cloth on which the paper is made.

Threads embodied in the web of the paper. Colored threads systematically arranged were formerly used in England for post-office envelopes and exchequer bills.

Silken fibers mixed with the pulp or dusted upon it in process of formation, as used in the United States currency.

Tigere, 1817, treated the pulp of the paper, previous to sizing, with a solution of prussiate of potash.

Sir Win. Congreve, 1819, prepared a colored layer of pulp in combination with white layers, also by printing upon one sheet and covering it with an outer layer, either plain or water-marked.

Glynn and Appel, 1821, mixed a copper salt in the pulp and afterward added an alkali or alkaline salt to produce a copious precipitate. The pulp was then washed and made into paper and thereafter dipped in a saponaceous compound.

Stevenson, 1837, incorporated into paper a metallic base such as manganese, and a neutral compound like prussiate of potash, to protect writing from being tampered with.

Varnham, 1845, invented a paper consisting of a white sheet or surface on one or both sides of a colored sheet.

Stones, 1851. An iodide or bromide in connection with ferrocyanide of potassium and starch combined with the pulp.

Johnson, 1853, employed the rough and irregular surface produced by the fracture of cast iron or other brittle metal to form a water mark for paper by taking an impression therefrom on soft metal, gutta- percha, etc., and afterward transferring it to the wire cloth on which the paper is made.

Scoutteten, 1853, treated paper with caoutchoue dissolved in bisulphide of carbon, in order to render it impermeable and to prevent erasures or chemical action.

Ross, 1854, invented water-lining or printing the denomination of the note in colors while the pulp was yet soft.

Evans, 1854, commingled a lace or open-work fabric in the pulp.

Courboulay, 1856, mixed the pulp and applied to the paper salts of iodine or bromine.

Loubatieres, 1857, manufactured paper in layers, any or all of which might be colored, or have impressions or conspicuous marks for preventing forgery.

Herapath, 1858, saturated paper during or after its manufacture with a solution of a ferrocyanide, a ferriccyanide, or sulphocyanide of potassium, sodium, or ammonium.

Seys and Brewer, 1858, applied aqueous solutions of ferrocyanide of potassium or other salts, which formed an indelible compound with the ferruginous base of writing ink.

Sparre, 1859, utilized opaque matter, such as prussian blue, white or red lead, insoluble in water and stenciled on one layer of the paper web, forming a regular pattern; this was then covered by a second layer of paper.

Moss, 1859, invented a coloring matter prepared from burned china or other clay, oxide of chromium or sulphur, and combined it with the pulp.

Barclay, 1859, incorporated with the paper:

1. Soluble ferrocyanides, ferricyanides, and sulphocyanides of various metals, by forming dibasic salts with potassium, sodium, or

ammonium, in conjunction with vegetable, animal, or metallic coloring matters.

2. Salts of manganese, lead, or nickel not containing ferrocyanogen.

3. Ferrocyanides, etc., of potassium, sodium, and ammonium, in conjunction with insoluble salts of manganese, lead, or nickel.

Hooper, 1860. Employed oxides of iron, either alone or dissolved in an acid, and mixed with the pulp.

Nissen, 1860. Treated paper with a preparation of iron, together with ammonia, prussiate of potash and chlorine, while in the pulp or being sized.

Middleton, 1860. Joined together one portion of a bank note printed upon one sheet of thin paper and the other part on another; the two were then cemented together by india-rubber, gutta-percha, or other compound. The interior printing could be seen through its covering sheet, so that the whole device on the note appeared on its face.

Olier, 1861. Employed several layers of paper of various materials and colors; the middle one was colored with a deleble dye, whose color was changed by the application of chemicals to the outer layer.

Olier, 1863. Prepared a paper of three layers of different thicknesses, the central one having an easily removable color, and the external layers were charged with silicate of magnesia or other salt.

Forster and Draper, 1864. Treating paper during or after manufacture with artificial ultramarine and Prussian blue or other metallic compound.

Hayward, 1864. Incorporated threads of fibrous materials of different colors or characters into and among the pulp.

Loewenberg, 1866. Introduced prussiate of potash and oxalic acid or such other alkaline salts or acids into the pulp, in order to indicate fraud in the removal of cancellation stamps or written marks.

Casilear, 1868. Printed numbers on a fugitive ground, tint or color in order to prevent alteration of figures or numbers.

Jameson, 1870. Printed on paper, designs with ferrocyanide of potassium and then soaked the paper when dry in a solution of oxalic acid in alcohol.

Duthie, 1872. Made a ground work of writing ink of different colors by any known means of pen ruling.

Syms, 1876. Produced graduated colored stains, which were made to partially penetrate and spread in the pulp web.

Van Nuys, 1878. Colored the Paper with a pigment and then printed designs with a soluble sulphide.

Casilear, 1878. United two distinctive colored papers, one a fugitive and the other a permanent color.

Hendrichs, 1879. Dipped ordinary paper in an aqueous solution of sulphate of copper and carbonate of ammonia and then added alkaline solutions of cochineal or equivalent coloring matter.

Nowlan, 1884. Backed the ordinary chemical paper with a thin sheet of waterproof paper.

Menzies, 1884. Introduced iodide and iodate of potassium or their equivalents into paper.

Clapp, 1884. Saturated paper with gallo-tanic acid, but the ink used on this paper contained ferri-sesquichloride or other similar preparation of iron.

Hill, 1885. Introduced into paper, ferrocyanide of manganese and hydrated peroxide of iron.

Schreiber, 1885. Colored paper material with indigo and with a subsequent treatment of chromates soluble only in alcohol.

Schreiber, 1885. Treated finished paper with ferric- oxide salts and with ferrocyanides insoluble in water but soluble in acids.

Schlumberger, 1890. Impregnated white paper with a resinated ferrous salt, a resin compound of plumbic ferrocyanide, and a resin compound of ferrocyanide of manganese in combination with a salt of molybdenum and a resin compound of zinc sulphide.

Schlumberger, 1893. Dyed first the splash fibers and mixed them with the paper pulp. Second. He also treated portions of the surface with an alkali, so as to form lines or characters thereon, then immersed the same in a weak acid, in order to produce water-mark lines.

Carvalho, 1894. 1. Charged the paper with bismuth iodide and sodium iodide. 2. Charged the paper with a bismuth salt and iodide of soda in combination with primulin, congo red or other pigment. 3. Charged the paper with a benzidine dye and an alkaline iodide.

1895. Applied a compound, sensitive to ink erasing chemicals, AFTER the writing has been placed on the paper.

Hoskins and Weis, 1895, a safety paper having added thereto a soluble ferrocyanide and a per-salt of iron insoluble in water but decomposable by a weak acid in the presence of a soluble ferrocyanide, as and for the purpose described. (2) A safety paper having added thereto a ferrocyanide soluble in water, a per-salt of iron insoluble in water but easily decomposed by weak acids in the presence of a ferrocyanide soluble in water, and a salt of manganese easily decomposed by alkalis or bleaching agents, substantially as described.

A review of the various processes for treatment of paper in pulp or when finished, demonstrates that time, money and study has been devoted to the production of a REAL safety paper. Some compositions and processes have in a measure been successful. It is found, however, that the ingenuity of those evil-minded persons, to the detection of whose efforts to alter the writing in documents this class of invention has more particularly been directed, finds a ready way of removing in some cases the evidence which the chemical reagent furnishes. This being true most of them have become obsolete, having entirely failed to accomplish the purposes for which they were invented.

There are but three so-called safety papers now on the market, if we exclude those possessing printed designs in fugitive colors.

It is a strange anomaly, nevertheless it is true, that 90 per cent or more of the "raised" checks, notes, or other monetary instruments which were in their original condition written on ordinary or so-called safety paper, never could have been successfully "put through" but for the gross and at times criminal negligence of their writers by the failure to adopt precautions of the very simplest kinds, and thereby avoided placing temptation in the way of many who under other circumstances would never have thought of becoming forgers.

There is no safety paper, safety ink, or mechanical appliance which will prevent the insertion of words or figures before other words or figures if a blank space be left where the forger can place them.

CHAPTER XXXII.

CURIOSA (INK AND OTHER WRITING MATERIALS).

ARTIFICIAL INK AND PAPER OWE THEIR INVENTION TO THE WASP—PHoeNICIA, "LAND OF THE PURPLE-DYE" —LINES, ADDRESSED TO THE PHoeNICIAN—OLDEST EXISTING PIECE OF LITERARY COMPOSITION—WHERE PAPYRUS STILL GROWS—DU CANGE'S LINES ON THE STYLUS—MATERIALS USED TO PROMULGATE ANCIENT LAWS OF GREECE—ANCIENT METHOD OF WRITING WILLS—MATERIALS EMPLOYED IN ANCIENT HEBREW ROLLS—ANTIQUITY OF EXISTING HEBREW WRITING —OLDEST SPECIMEN OF GREEK WAX WRITING—WOODEN TALLIES AS EMPLOYED IN ENGLAND—WHEN WRITING IN GOLD CEASED—DATE OF THE FIRST DISCOVERY OF GREEK PAPYRUS IN EGYPT—PERIODS TO WHICH BELONG VARIOUS STYLES OF WRITING—ANECDOTE AND POEM ABOUT THE FIRST GOLD PEN—INTERESTING NOTES ABOUT PENS AND INK-HORNS—EMPLOYMENT OF THE PEN AS A BADGE IN THE FOURTEENTH CENTURY—SOME LINES BY COCKER—THE OLDEST EXISTING WRITTEN DOCUMENTS OF RUSSIA—WHEN SEALING WAX WAS FIRST EMPLOYED—PLINY'S DESCRIPTION OF THE DIFFERENT KINDS OF PAPYRUS PAPER—MODE OF PRESERVING THE ANCIENT PAPYRUS ROLLS—SUGGESTIONS RESPECTING USES OF INK—COMPARATIVE TABLE ABOUT COAL TAR AND ITS BY-

PRODUCTS—COMPOSITIONS OF SECRET INKS AND HOW TO RENDER THEM VISIBLE—CHARACTER OF INK EMPLOYED FOR MANY YEARS BY THE WASHINGTON PATENT OFFICE—FACTS ELICITED BY HERAPATH IN THE UNROLL-MENT
OF A MUMMY—LINES FROM SHAKESPEARE AND PERSEUS—SEVENTEENTH CENTURY OBSERVATIONS ABOUT SECRET INKS—CAUSE OF THE DESTRUCTION OF MANY ANCIENT MSS.—METHODS TO BE EMPLOYED IN THE RESTORATION OF SOME OLD INKS—VARIATIONS IN THE MEANING OF WORDS—THE POUNCE BOX PRECEDED BLOTTING PAPER—SOME OBSERVATIONS ABOUT BLOTTING PAPER—ANECDOTE RELATING TO DR. GALE—WHEN WAFERS WERE INTRODUCED—PERSIAN ANECDOTE ABOUT THE DIVES—EPISODES RESPECTING THE STYLUS—DESCRIPTION BY BELOE OF ANCIENT PERSIC AND ARABIC MSS.—CITATION FROM OLD BOSTON NEWSPAPER AND POEM—METHOD OF COLLECTING RAGS IN 1807 AND SOME LINES ADDRESSED TO THE LADIES—METHOD TO PHOTOGRAPH COLORED INKS—POEM BY ISABELLE HOWE FISKE.

IN considering the important and kindred subjects of "gall" ink and "pulp" paper, we are not to forget the LITTLE things connected with their development and which, indeed, made their invention possible.

The gall-nut contains gallic and gallo-tannic acid, and which acids, in conjunction with an iron salt, forms the sole base of the best ink. This nut is produced by the punctures made on the young buds of branches of certain species of oak trees by the female wasp. This same busy little insect was also the first professional paper maker. She it was who taught us not only the way to change dry wood into a suitable pulp, the kind of size to be used, how to waterproof and give the paper strength, but many more marvelous details appertaining to the manufacture of paper which in their ramifications have proved of inestimable benefit and service to the human race. * * * * * * *

The Greek word "Phoenicia" means literally "the land of the purple dye," and to the Phoenicians is attributed the invention of the art of writing.

TO THE PHOENICIAN.

"Creator of celestial arts,
Thy painted word speaks to the eye;
To simple lines thy skill imparts
The glowing spirit's ecstasy."

The oldest piece of literary composition known in the oldest book (roll) in existence is to be found in the celebrated papyrus Prisse, now in the Louvre at Paris. It consists of eighteen pieces in Egyptian hieratic writing, ascribed to about the year B. C. 2500.

While the papyrus plant has almost vanished from Egypt, it still grows in Nubia and Abyssinia. It is related by the Arab traveler, Ibn-Haukal, that in the tenth century, in the neighborhood of Palermo in Sicily, the papyrus plant grew with luxuriance in the Papirito, a stream to which it gave its name.

Du Cange, 1376, cites the following lines from a French metrical romance, written about that time, to show that waxen tablets continued to be occasionally used till a late period:

"Some with antiquated style
In waxen tablets promptly write;
Others with finer pen, the while
Form letters lovelier to the sight."

The laws of Greece were promulgated by means of MSS. on linen, as they were also in Rome, and in addition to linen; cloth and silk were occasionally used. Skins of various kinds of fish, and even the "intestines of serpents" were employed as writing materials. Zonaras states that the fire which took place at Constantinople in the reign of Emperor Basiliscus consumed, among other valuable remains of antiquity, a copy of the Iliad and Odyssey, and some other ancient poems, written in letters of gold upon material formed of the intestines of a serpent. We are also informed by Purcelli that monuments of much more modern dates, the charter of Hugo and Lothaire, A. D. 933 (kings of Italy), preserved in the archives of Milan, are written upon fish skins.

Constantine authorized his soldiers dying on the field of battle to write their last will and testament with the point of their sword on its sheath or on a shield.

B. C. 270. The Jewish elders, by order of the high priest, carried a copy of the law to Ptolemy Philadelphus, written in letters of gold upon skins, the pieces of which were so artfully put together that the joinings did not appear.

No monuments of Hebrew writing exist which are not posterior even to the Christian era, with the exception of those on the coins of the Maccabees, which are in the ancient or what is termed the Samaritan forms of the Hebrew letters. This coinage took place about B. C. 144.

The most ancient specimen of Hebrew ink writing extant is alleged to have been written A. D. 489. It is a parchment roll which was found in a Kariat synagogue in the Crimea. Another, brought from Danganstan, if the superscription be genuine, has a date corresponding with A. D. 580. The date of still another of the celebrated Hebrew scriptural codices, about which there is no dispute, is the Hilel codex written at the end of the sixth century. Its name is said to be derived from the fact that it was written at Hila, a town built near the ruins of the ancient Babel; some maintain, however, that it was named after the man who wrote it.

One of the earliest specimens of Greek (wax) writing is an inscription on a small wooden tablet now in the British museum. It refers to a money transaction of the thirty-first year of Ptolemy Philadelphus, B. C. 254.

In England the custom of using wooden tallies, inscribed as well as notched in the public accounts, lasted down to the nineteenth century.

Gold writing was a practice which died out in the thirteenth century.

The first discovery of Greek papyri in Egypt took place in the year 1778. It is of the (late of A. D. 191 and outside of Egypt and Herculaneum is the only place in which the Greek papyri has ever been found.

Square capital ink writing in Latin of ancient date is found on a few leaves of an MS. of Virgil, which is attributed to the close of the fourth century, and the first rustic MS. to which an approximate date can be given, belongs to the close of the fifth century.

The most ancient uncial ink writing extant, belongs to the fourth century, whilst the earliest mixed uncial and miniscule writing pertains to the sixth century.

The oldest extant Irish MS. in the round Irish hand is ascribed to the latter part of the seventh century, while the earliest specimen of English writing of any kind extant dates about the beginning of the eighth century.

The gold pen won by Peter Bales in his trial of skill with Johnson, during the reign of Queen Elizabeth, if really made for use, is probably the first modern example of such pens. Bales was employed by Sir Francis Walsingham, and afterwards kept a writing school at the upper end of the Old Bailey. In 1595, when nearly fifty years old, he had a trial of skill with one Daniel Johnson, by which he was the winner of a golden pen, of a value of L20, which, in the pride of his victory, he set up as his sign. Upon this occasion John Davis made the following epigram in his "Scourge of Folly:"

"The Hand and Golden Pen, Clophonion
Sets on his sign, to shew, O proud, poor soul,
Both where he wonnes, and how the same he won,
From writers fair, though he writ ever foul;
But by that Hand, that Pen so borne has been,
From place to Place, that for the last half Yeare,
It scarce a sen'night at a place is seen.
That Hand so plies the Pen, though ne'er the neare,
For when Men seek it, elsewhere it is sent,
Or there shut up as for the Plague or Rent,
Without which stay, it never still could stand,
Because the Pen is for a Running Hand."

The sign of the "Hand and Pen" was also used by the Fleet street marriage-mongers, to denote "marriages performed without imposition."

Robert More, a famous writing master, in 1696 lived in Castle street, near St. Paul's churchyard, London, at the sign of the "Golden Pen."

The ink horn in Queen Elizabeth's time was in popular use as a receptacle for holding writing ink, and Petticoat lane in London was the great manufacturing center for them. Bishops Gate in the same vicinity was known as the "home of the scribblers."

Beginning with 1560 and for many years thereafter the sign of the Five Ink Horns was appropriately displayed by Haddon on the house in which he dwelt.

Away back in the time of King Edward III (1313- 1377), royalty was employing the pen, both quill and gold, as badges. This is indicated in the accompanying interesting list to be found in the Harlein library:

"King Edward the iii. gave a lyon in his proper coulor, armed, azure, langue d'or. The oustrich fether gold, the pen gold, and a faucon in his proper coulor and the Sonne Rising.

"The Prince of Wales the ostrich fether pen and all arg.

"Henry, sonne of the Erl of Derby, first Duk of Lancaster, gave the red rose uncrowned, and his ancestors gave the Fox tayle in his prop. coulor and the ostrich fether ar. the pen ermyn.

"The Ostrych fether silver, the pen gobone sylver and azur, is the Duk of Somerset's bage.

"The ostrych fether silver and pen gold ys the kinges.

"The ostrych fether pen and all sylver ys the
Prynces.

"The ostrych fether sylver, pen ermyn is the
Duke of Lancesters.

"The ostrych fether sylver and pen gobone is
the Duke of Somersets."

"What's great Goliath's spear, the sevenfold shield,
Scanderbeg's sword, to one who cannot wield
Such weapons? Or, what means a well cut quill,
In th' untaught hand of him that's void of skill?"
—COCKER, A. D. 1650.

The oldest ink (Russian) documents that exist in Russia are two treaties with the Greek emperors, made by Oleg, A. D. 912, and Igor, A. D. 943. Christianity, introduced into Russia at the beginning of the eleventh century by Vladimir the Great, brought with it many words of Greek origin. Printing was introduced there about the middle of the sixteenth century. The oldest printed book which has been discovered is a Sclavonic psalter, the date Kiev, 1551, two years after a press was established in Moscow.

It is said that the skins of 300 sheep were used in every copy of the first printed Bible. Hence the old saying, "It takes a flock of sheep to write a book."

What would have been the comment in olden times, to learn that it takes almost a forest of trees to print the Sunday edition of some of our great newspapers?

Wax (shoemakers') was first employed on documents A. D. 1213, although it was white wax which was used to seal the magna charta, granted to the English barons by King John, A. D. 1215. In 1445 red wax was much employed in England, but the earliest specimen of red sealing wax extant is found on a letter dated August 3, 1554.

Pliny enumerates and describes eight different kinds of papyrus paper:

1. Charter hieratica—sacred paper, used only for books on religion. From adulation of Augustus it was also called charta augusta and charta livia.

2. Charta amphitheatrica—from the place where it was fabricated.

3. Charta fannia—from Fannius, the manufacturer.

4. Charta saitica—from Sais in Egypt. This appears to have been a coarser kind.

5. Charta toeniotica—from the place where made, now Damietta. This was also of a less fine quality.

6. Charta claudia. This was an improvement of the charta hieratica, which was too fine.

7. Charta emporitica. A coarse paper for parcels.

There was also a paper called macrocollum, which was of a very large size.

Of all these, he says, the charta claudia was the best.

The ink-written rolls of papyrus were placed vertically in a cylindrical box called capsula. It is very evident that a great number of such volumes might be comprised in this way within a small space, and this may tend to explain the smallness of the rooms which are considered to have been used for containing the ancient libraries.

At Mentz, in Upper Germany, is a leaf of parchment on which are fairly written twelve different kinds of handwritings in six different inks also a variety of miniatures and drawings curiously done with a pen by one Theodore Schubiker, who was born without hands and performed the work with his feet.

In Rome the very plate of brass on which the laws of the ten tables are written is still to be seen.

Stylographic inks should not be used upon records, most of them are aniline. The absence of solid matter, which makes them desirable for the stylographic pen, unfits them for records.

Never add water to ink. While an ink which has water as its base might, under certain conditions bear the addition of an amount equal to that lost by evaporation, as a rule the ink particles which have become injured will not assimilate again.

One of the best methods to cleanse a steel pen after use, is to stick it in a raw (white) potato.

Inks which are recommended as permanent, because water will not remove them, while it does immediately obliterate others, may not be permanent as against time. These inks may be the best for monetary purposes, but, owing to an excess of acid in them, may be dangerous in time to the paper.

It is interesting, since coal tar has acquired so important a position in the arts, to trace how its various products successively rose in value. The prices in Paris, as given by M. Parisal in 1861, are as follows:

Coal,.................................. 1/4 c. per lb.
Coal tar,.............................. 3/4 " "
Heavy coal oil,............. 2 1/2 a 3 3/4 " "
Light coal oil,............ 6 3/4 a 10 1 /4 " "
Benzole,....................... 10 1/2 a 13 " "
Crude nitro-benzole,................ 57 a 61 " "
Rectified nitro-benzole,............ 82 a 96 " "
Ordinary aniline,............ $3.27 a $4.90 " "
Liquid aniline violet,............. 28 a 41 " "
Carmine aniline violet,....... 32 c. a $1.92 "
Pure aniline violet, in powder,.... $245 a $326.88 "

The last is equal to the price of gold. And so, says M. Parisal, from coal, carried to its tenth power, we have gold; the diamond is to come.

Modern chemistry offers many formulas and methods of rendering visible secret or sympathetic inks. Writing made with any of the following solutions, and permitted to dry, is invisible. Treatment by the means cited will render them visible.

Solution. After treatment. Color produced.
Acetate of lead. Sulphuret of potassiurin. Brown.
Gold in nitrohydroChloric acid. Tin in same acid. Purple.
Nut-galls. Sulphate of iron. Black.
Dilute sulphuric acid. Heat. Black.
Cobalt in dilute Heat. Green.
nitrohydrochloric acid.
Lemon juice. Heat. Brown.
Oxide of copper in Heat. Blue.
acetic acid and salt
Nitrate of bismuth. Infusions of Nutgalls. Brown.
Common starch. Iodine in alcohol. Purple.

Colorless iodine. Chloride of lime. Brown.
Phenolphtalin. Alkaline solution. Red.
Vanadium. Pyrogallic acid. Purple.

The Patent Office at Washington, D. C., for more than forty years employed a violet copying ink made of logwood. From 1853 until 1878 it was furnished by the Antoines of Paris, of the brand termed "Imperial;" in later years it was supplied by the Fabers. Since 1896 they have been using "combined" writing fluids.

The following facts elicited by the unrollment of a mummy at Bristol, England, in 1853, were communicated to the Philosophical Magazine, by Dr. Herapath. He says:

"On three of the bandages were hieroglyphical characters of a dark color, as well defined as if written with a modern pen; where the marking fluid had flowed more copiously than the characters required, the texture of the cloth had become decomposed and small holes had resulted. I have no doubt that the bandages were genuine, and had not been disturbed or unfolded; the color of the marks were so similar to those of the present 'marking ink,' that I was induced to try if they were produced by silver. With the blow-pipe I immediately obtained a button of that metal; the fibre of the linen I proved by the microscope, and by chemical reagents, to be linen; it is therefore certain that the ancient Egyptians were acquainted with the means of dissolving silver, and of applying it as a permanent ink; but what was their solvent? I know of none that would act on the metal and decompose flax fibre but nitric acid, which we have been told was unknown until discovered by the alchemist in the thirteenth century, which was about 2200 years after the date of this mummy, according as its superscription was read.

"The Yellow color of the fine linen cloths which had not been stained by the embalming materials, I found to be the natural coloring matter of the flax; they therefore did not, if we judge from this specimen, practice bleaching. There were, in some of the bandages near the selvage, some twenty or thirty blue threads; these were dyed by indigo, but the tint was not so deep nor so equal as the work of the modern dyers; the color had been given it in the skein.

"One of the outer bandages was of a reddish color, which dye I found to be vegetable, but could not individualize it; Mr. T. J. Herapath analyzed it for tin and alumina, but could not find any. The face and internal surfaces of the orbits had been painted white, which pigment I ascertained to be finely powdered chalk."

"I am a scribbled form, drawn with a Pen
Upon a Parchment, and against this fire
Do I shrink up."
—KING JOHN, v, 7.

"With much ado, his Book before him laid,
And Parchment with the smoother side display'd;
He takes the Papers, lays 'em down agen,
And with unwilling fingers tries his Pen;
Some peevish quarrel straight he tries to pick,
His Quill writes double, or his Ink's too thick;
Infuse more Water; now 'tis grown too thin,
It sinks, nor can the characters be seen."
—Persius, translated by Dryden.

INKS CALLED SYMPATHETICAL (Seventeenth Century).

"These operations are liquors of a different nature, which do destroy one another; the first is an infusion of quick-lime and orpin; the second a water turn'd black by means of burned cork; and the third is a vinegar impregnated with saturn.

"Take an ounce of quick-lime, and half an ounce of orpin, powder and mix them, put your mixture into a matrass, and pour upon it five or six ounces of water, that the water may be three fingers breadth above the powder, stop your matrass with cork, wax, and a bladder; set it in digestion in a mild sand heat ten or twelve hours, shaking the matrass from time to time, then let it settle, the liquid becomes clear like common water.

"Burn cork, and quench it in aqua vitae, then dissolve it in a sufficient quantity of water, wherein you shall have melted a little gumm arabick, in order to make an ink as black as common ink. You must separate the cork that can't dissolve, and if the ink be not black enough, add more cork as before.

"Get the impregnation of saturn made with vinegar, distilled as I have shewn before, or else dissolve so much salt of saturn as a quantity of water is able to receive: write on paper with a new pen dipt in this liquor, take notice of the place where you writ, and let it dry, nothing at all will appear.

"Write upon the invisible writing with the ink made of burnt cork, and let it dry, that which you have writ will appear as if it had been done with common ink.

"Dip a little cotton in the first liquor made of lime and orpin, but the liquor must be first settled and clear; rub the place you writ upon with this cotton and that which appeared will presently disappear, and that which was not seen will appear.

ANOTHER EXPERIMENT.

Take a book four fingers breadth in bigness, or bigger if you will: write on the first leaf with your impregnation of saturn, or else put a paper that you have writ upon between the leaves; turn to t' other side of the Book, and having observed as near as may be the opposite place to your writing, rub the last leaf of the book with cotton dipt in liquor made of quick- lime and orpin, nay and leave the cotton on the place clap a folded paper presently upon it, and shutting the book quickly, strike upon it with your hand four or five good strokes; then turn the book, and clap it into a press for half a quarter of an hour; take it out and open it, you'll find the place appear black, where you had writ with the invisible ink. The same thing might be done through a wall, if you could provide something to lay on both sides, that might hinder the evaporation of the spirits.

REMARKS.

"These operations are indeed of no use, but because they are somewhat surprizing, I hope the curious will not take it ill, that I make this small digression.

"It is a hard matter to explicate well the effects I have now related, nevertheless I shall endeavour to illustrate them a little, without having recourse to sympathy and antipathy, which are general terms, and do not explicate nothing at all; but before I begin, we must remark several things.

"The first is, that it is an essential point to quench the coal of cork in aqua vitae, that the visible ink may become black with it.

"Secondly, that the blackness of this ink does proceed from the fuliginosity or sooty part of the coal of the cork which is exceeding porous and light, and that this fuliginosity is nothing but an oil very much rarefied.

"Thirdly, that the impregnation of saturn, which makes the invisible ink, is only a lead dissolved, and held up imperceptibly in an acid liquor, as I have said, when I spoke of this metal.

"Fourthly, that the first of these liquors in a mixture of the alkali and igneous parts of quick-lime with the sulphureous substance of arsenick; for the orpin is a sort of arsenick, as I said before.

"All this being granted, as no body can reasonably think otherwise, I now affirm, that the reason why the visible ink does disappear, when the defacing liquor is rubbed upon it, is that this liquor consisting of an alkali salt, and parts that are oily and penetrating, this mixture does make a kind of soap, which is able to dissolve any fuliginous substance, such as burnt cork, especially when it has been already rarefied and disposed for dissolution by aqua vitae, after the same manner as common soap, which is compounded of oil, and an alkali salt, is able to take away any spots made by grease.

"But it may be demanded, why after the dissolution the blackness does disappear.

"I answer, that the fuliginous parts have been so divided, and locked up in the sulphureous alkali of the liquor, that they are become invisible, and we see every day that very exact solutions do render the thing dissolved imperceptible, and without colour.

"The little alkali salt which is in the burnt cork may also the better serve to joyn with the alkali of the quick-lime, and to help the dissolution.

"As for the invisible ink, it is easy to apprehend how that appears black, when the same liquor, which serves to deface the other, is used upon it. For whereas the impregnation of saturn is only a lead suspended by the edges, of the acid liquor, this lead must needs revive, and resume its black colour, when that which held it rarefied is entirely destroyed; so the alkali of quick- lime being filled with the sulphurs of arsenick becomes very proper to break and destroy the acids, and to agglutinate together the particles of lead.

It happens that the visible ink does disappear by reason that the parts which did render it black have been dissolved; and the invisible ink does also appear because the dissolved parts have been revived.

"Quick-lime and, orpiment being mixed and digested together in water, do yield a smell much like that which happens when common sulphur is boiled in a lixivium, of tartar. This here is the stronger, because the sulphur of arsenick is loaded with certain salts that make a stronger impression on the smell. Quick- lime is an alkali that operates in this much like the salt of tartar in the other operation; you must not leave the matrass open, because the force of this water doth consist in a volatile.

"The lime retains the more fixt part of the arsenick and the sulphurs that come forth are so much the more subtile, as they are separated from what did fix them before, and this appears to be so, because the sulphurs must of necessity pass through all the book to make a writing of a clear and invisible liquor appear black and visible: and to facilitate this penetration the book is strook, and then turned about, because the spirit or volatile sulphurs do always tend upwards; you must likewise clap it into a press, that these sulphurs may not be dispersed in the air. I have found, if that these circumstances are not observed, the business fails. Furthermore that which persuades me that the sulphurs do pass through the book, and not take a circuit to slip in by the sides, as many do imagine, is that after the book is taken out of the press, all the inside is found to be scented with the smell of this liquor.

"There is one thing more to be observed, which is, that the infusion of quick-lime and orpin be newly made, because otherwise it will not have force enough to penetrate. The three liquors should be made in different places too; for if they should approach near one another, they would be spoiled.

"This last effect does likewise proceed from the defacing liquor; for because upon the digestion of quick- lime and orpin, it is a thing impossible for some of the particles will exalt, stop the vessel as close as you will; the air impregnated with these little bodies does mix with, and alter the inks, insomuch that the visible ink does thereby become the less black, and the invisible ink does also acquire a little blackness."

Priceless MSS. in immense number written in periods between the third and thirteenth centuries have been destroyed by modern scholars in experimentations based on the false theory that the faded inks on them, whether above or below other inks (palimpsests), contained iron.

Sulphocyanide of potassium is highly esteemed as a reagent for the restoration of writing, if iron is present. Theoretically, it is one of the best for such a purpose if employed with acetic acid. It causes, however, such a decided contraction of parchment as to be utterly useless, but for paper MSS. is excellent. The metallic sulphides generally pronounced harmless, causes the writing to soften and become illegible in a short time. On the other hand, yellow prussiate of potash, with acetic acid in successive operations is of great service in treating the most perplexing palimpsests.

Ink which badly corrodes a steel pen need not necessarily be condemned; it may contain just the qualities which make it bind to the paper and render it more durable.

Some inks which are fairly permanent against time
if not tampered with, can be removed with water.
This is true of the most lasting of inks,—the old
"Indian."

In ancient Latin MSS. the words fuco, fucosus and fucus are found to be frequently employed. It is interesting to note the variations in their meaning:

FUCO.—To color, paint or dye a red color.

FUCOSUS.—Colored, counterfeit, spurious, painted, etc.

FUCUS.—Rock lichen (orchil) red dye. Red or purple color. The (reddish) juice with which bees stop up the entrance to their hives. Bee glue.

FUCUS.—A drone.

In Japan the word "ink" possesses more than one meaning Four hundred Inks—one degree of sixty miles." (See Geographical Grammar, of 1737, page 3.)

"Say what you will Sir, but I know what I know;
That you beat me at the Mart, I have your hand to show;
If the skin were Parchment, and the blows you gave were Ink,
Your own Hand-writing would tell you what I think."
—Comedy of Errors, iii, 1.

The first book ever printed in Europe, to wit, a copy of "Tully's Offices," is carefully preserved in Holland.

White's Latin-English Dictionary, 1872, distinguishes the words Atramentum and Sutorium in their interpretations.

ATRAMENTUM.—The thing serving for making black. A black liquid of any kind. A writing ink. Shoemaker's black. Blue vitriol.

SUTORIUM.—Belonging to a shoemaker.

Before the employment of blotting paper a pounce- box which contained either powdered gum sandarach and ground cuttle-fish bones, or powdered charcoal, sand and like materials was used by shaking it like a pepper- box on freshly written manuscripts.

Blotting paper as first employed consisted of very thin sheets and of a dark pink color, which fashion changed to blue in later years.

Good blotting paper of the present time removes fully two thirds of fresh ink when used on HARD finished paper.

Blotting paper should not be used upon records. Its use removes the body of the ink, leaving discoloration, but nothing for penetration. In inks intended for copying, the employment of blotting paper is especially bad.

"Thou hast most traitorously corrupted the youth of the realm in erecting a Grammar School; and whereas, before, our forefathers had no other books but the score and the tally, thou hast caused printing to be used, and contrary to the King, his crown and dignity, thou hast built a paper mill." —2 King Henry VI, iv, 5.

Mr. Knight relates a conversation between Dr. Gale and a gentlemen from the West relative to the introduction of some material into ink to prevent moulding. Dr. Gale had astonished his friend by stating— "will prevent the deposition of the ova of infusoria animalcutae;" when it was suggested that he add "and the sporadic growths of thallogenic cryptograms and be fatal to the fungi."

The University of Pennsylvania claims to possess the oldest piece of writing in the world and which is on a fragment of a vase found at Nippur. It is an inscription in picture writing supposed to have been made 4,500 years before Christ.

Wafers were not introduced until the close of the sixteenth century.

The Persians in ancient times, some 800 years B. C., were in the habit of celebrating certain festivals and it is related that in the month of December one of their ceremonies was that of driving the Dives (spirits) out of their houses.

For this purpose the Magi wrote certain words with saffron on skins, papyrus or wood and then smoked it over a fire. The spell thus prepared was glued or nailed to the inside of the door, which was painted red. The priest then took sand, which he spread with a long knife, whilst he muttered certain prayers and then throwing it on the floor the enchantment was complete; and the Dives were supposed immediately to vanish; or at least to be deprived of all malignant influence.

Aristotle's work on the Constitution of Athens, B. C. 340, or probably the copy made by Tyrannio, was discovered transcribed underneath farm accounts of land in the district of Hermopolis in Egypt in the reign of Vespasian, A. D. 9 to 79.

In MSS. written before the invention of printing and indeed for many years after, the title page if any, will be found on the last page with the date.

"Let lawyers bawl and strain their throats,
'Tis I that must the lands convey,
And strip their clients to their coats,
Nay, give their very souls away!"
—DEAN SWIFT, "On ink."

"It is certain that in their treaties with the European Greeks of Constantinople the Arabs always stipulated for the delivery of a fixed number of manuscripts. Their enthusiasm for Aristotle is equally notorious; but it would be unjust to imagine that, in adopting the Aristotelian method, together with the astrology and alchemy of Persia, and of the Jews of Mesopotamia and Arabia, they were wholly devoid of originality."

The "Arabic" numerals which we now employ are probably of Indian origin, having been brought by Arab traders from the East and introduced by them into Spain in the middle ages, whereas they spread over Europe coming in use in England perhaps about the eleventh century. But whether India invented them or borrowed from Greek or other traders from the West is unknown.

The ancient writing implement known as the stylus was made of every conceivable material, sometimes with the precious metals, but usually of iron, and on occasion might be turned into formidable weapons. It was with his stylus that Caesar stabbed Casca in the arm, when attacked in the senate by his murderers; and Caligula employed some person to put to death a senator with a like instrument.

In the reign of Claudius women and boys were searched to ascertain whether there were any styluses in their pen cases. Stabbing with the pen, therefore, is not merely a metaphorical expression.

Sir William Gore Ouseley, a famous diplomat and savant, who was living at the beginning of the nineteenth century, during his long residence in India spent a fortune in the collection of ancient Persic and Arabic MSS. In 1807 he permitted them to be examined by Beloe, whose description of a few will bear repeating:

"No. 1. A Koran, in the Cufi or Cufic character, said to be written by Ali, the son-in-law of Mahammed, the Arabian prophet. The substance upon which this curious manuscript is written appears to be a fine kind of asses' skin or vellum, and the ink of a red, brownish colour. The ends of verses are marked by large stars of gold. If written by Ali, it must be nearly twelve hundred years old, but at all events may be considered as very ancient, many hundred years having elapsed since the use of the Cufi character has given way to the Neskh, Suls, etc., etc. This manuscript is still in excellent preservation."

"No. 4. Beharistan, 'The Garden of Spring.' A book on ethics and education, illustrated by interesting anecdotes and narratives, written both in verse and prose, in imitation of the Gulistan, or 'Rose garden' of Saadi, and like it divided into eight chapters, composed by Nuruddin, Abdurrahman Jami, ben Ahmed of the village of Jam, near Herat. He was born A. H. 817 and died at the age of 81 years (about A. D. 1492). As a grammarian, theologist and poet he was unequalled, and his compositious are as voluminous as they are excellent. The enormous expense which people have incurred to possess accurate copies of and to adorn and embellish his works, is no small proof of the great estimation in which they were held by the literati of the East."

"This volume is a small folio, consisting of 134 pages, written in the most beautiful Nastilik character, by the famous scribe Mohammed Hussein, who, in consequence of his inimitable penmanship, obtained the title of Zerin Kalm, or 'Pen of Gold.' The leaves are of the softest Cashmirian paper, and of such modest shades of green, blue, brown, dove, and fawn colors, as never to offend the eye by their glare, although richly powdered with gold. The margins, which are broad, display a great variety of chaste and beautiful delineations in liquid gold, no two pages being alike. Some are divided into compartments, others are in running patterns, in all of which the illuminations show the most correct, and at the same time fanciful taste. Many are delineations of field sports, which, though simple outlines of gold, are calculated to afford the highest gratifications to the lover of natural history, as well as the artist, from the uncommon accuracy with which the forms of the elephant, rhinoceros, buffalo, lion, tiger, leopard, panther, lynx, and other Asiatic animals are portrayed. It appears, by the names which are inserted at the bottom of the pages, that several artists were employed in the composition and combination of these ornaments, one for the landscape, another for the animals, and a third for the human figures, all of whom have given proofs of superior merit. It would take almost a month to inspect all the excellencies of this rare manuscript; for, although so richly ornamented in gold, the chaste colors of the ground prevent any glaring obtrusion on the eye, and oblige the examiner to place it in a particular point of light to see the exquisite and minute beauties of the delineations. The paintings, which are meant to illustrate the subject of the book, are done in colors, and in the center of the leaves.

"On the back of the first page are the autographs of the Emperors of Hindustan, Jehangir and his son Shajehan."

"No. 5. 'A Diwan i Shahi.' A Diwan or Collection Odes by Shahi,' transcribed by the famous penman Mir Ali, in Bokar<a1., A. D. 1534. (A. H. 940.)

"The author of these poems, Mamlic Arnir Shahi, the son of Malic Jemaluddin Firozkohi, a nobleman of high rank and fortune as well as great literary attainments, was born in Sebzwar, A. H. 786. He passed a part of his life at the courts of Baisankar (the son of Shahrukh Mirza, and grandson of Tamerlane) and of his son Abul Kasim Baber, during which time he held appointments of the highest trust and emolument, and was universally caressed. But, taking offense at an expression of Sultan Baber's, which he conceived reflected on his father, he quitted the court in disgust, and passed the remainder of his life in the cultivation of the sister arts, poetry, painting, and music in all of which he eminently excelled. He was also unequalled in penmanship. At the age of seventy years be died in Asterabad, during the reign of Baber, A. H. 856, and was buried in the suburbs of his native city, Sebzwar, in a mausoleum erected by his ancestors.

"Mir Ali, who transcribed this book, was the most excellent penman of his time. He was born in the reign of Sultan Hussein Mirza Bahudur, the son of Mansur, and great grandson of Omar Sheikh, the second son of Tamerlane. He was a learned man and good poet, and took the Takhulas (poetical title) most appropriate to his greatest accomplishments, of Al Cateb, or 'the Scribe.' He was the pupil of Sultan Ali, but far exceeded his master in calligraphy. An entire book written by him is justly esteemed a great treasure in the East.

"On the back of the first page of this most beautiful manuscript are the autographs of the Emperors of Hindustan, Jehangir (the son of the great Acber) and his son Shah Jehan; there is also the seal of Aurangzeb, the son of Shah Jehun. Jehangir dates the acquiring possession of this treasure A. H. 1025, and Shah Jehun, A. H. 1037.

"A collection of mythological drawings (brought from a fort in Bhutan, where they were taken as plunder) exceedingly well coloured, and richly illumined. Some of the deities resemble those of the Tartars, delineated by the traveller Pallas; others again are pure Hindu and many Chinese; but the most frequent are the representations of Baudh, exactly as depicted in the paintings and temples at Ceylon. The religion of Bhutan and Neipal seems to be like the local situation of those countries, the link of connection between that of the Hindus, with its different schisms, and that of the Chinese with the Tartar superstructure.

"With this book of drawings are several rolls of Bhutan Scripture, very well stamped by stereotype blocks of wood. Some of the blocks accompanied the drawings; they are sharply and neatly cut in a kind of Sanscrit character, and are objects of great curiosity, as, by the accounts of the natives, this mode of printing has been in use for time immemorial."

"There are besides in Sir Gore Ouseley's collection 1,100 most beautiful books of Persian and Indian paintings, portraits of the Emperors of Hindustan from Sultan Baber down to Bahudur Shah, finely colored drawings of natural history, and curious designs of fancy, with specimens of fine penmanship in the different kinds of Arabic and Persian characters. Several Sanscrit manuscripts, highly ornamented and richly illumined, some of them written in letters of gold and silver on a black ground. Many of them illustrated with the neatest miniature paintings of the Hindu gods and saints. Two Korans, the letters entirely of gold, with the vowel points in black. The two versions of Pilpais or Bedpai's fables, by Hussein Vaiz and Abulfazl, illustrated with upwards of 700 highly finished miniatures; the best historical works in the Persian language, finely written, and in high preservation."

The high regard with which the writers of MSS. in ancient Persia were viewed may be learned among other things from the following anecdote:

One of the most eminent among them was in his walks solicited by a beggar for alms. "Money," he replied, "I have none," but taking his pen and ink from his girdle, which are the insignia of the profession (without which they never went abroad), he took a piece of paper, and wrote some word or other upon it. The poor man received it with gratitude, and sold it to the first wealthy person he met for a golden mohur, in value about $2.50.

"Is not this a lamentable thing, that of the skin of an innocent lamb should be made Parchment? that Parchment being scribbled o'er should undo a man?" —2 King Henry VI, iv, 2.

The Boston News Letter, 1769, announces:

"The belleart will go through Boston before the end of next month, to collect rags for the paper mill at Milton, when all people that will encourage the paper manufactory may dispose of them."

"Rags are as beauties, which concealed lie,
But when in paper how it charms the eye;
Pray save your rags, new beauties it discover,
For paper truly every one's a lover:

By the pen and press such knowledge is displayed,
As wouldn't exist, if paper was not made.
Wisdom of things, mysterious, divine,
Illustriously doth on paper shine."

Gen. Walter Martin, proprietor of the township of Martinsburg, Lewis county, N. Y., erected a paper-mill, which was run by John Clark & Co. This was in 1807. They gave notice that rags would be received at the principal stores in Upper Canada and the Black river country, which (like many of the advertisements of the early papermakers, both in England and America), was accompanied by a poetic address to the ladies, one stanza of which ran thus:

Sweet ladies pray be not offended,
Nor mind the jests of sneering wags;

No harm, believe us, is intended,
When humbly we request your rags."

The employment of complementary color screens has made it possible to photograph colors which formerly indicated no contrast with white back grounds in the negative and later in the finished picture.

This discovery has destroyed the value of "safety" papers, based on complete tints or possessing colored lines or words.

"IN MANUSCRIPT.

"The rain storm wields a noisy pen
  Adown the pane,
Wet splashes leaving, blots of strange white ink,
  Blunders of rain.

"And yet no poems of ecstatic men,
  Olympic faced,
Could be as wonderful as these, I think,
  In cipher traced."

—ISABELLE HOWE FISKE.